ADVANCED TACTICS IN AMERICA

Other Tactics/Intel. Manuals from Posterity Press:

Okinawan Reckoning
Peleliu Progress
The Iwo Alamo
Super-Squad: The Now Missing Component
North Korea's Hidden Assets
Guardian Joe: How Less Force Helps the Warrior
Sinoland: The Subversion of Freedom's Bastion
Afrique: A Warning for America
Strategic Rifleman: Key to More Moral Warfare
Gung Ho: The Corps' Most Progressive Tradition
Global Warrior: Averting WWIII
Expeditionary Eagles: Outmaneuvering the Taliban
Homeland Siege: Tactics for Police and Military
Tequila Junction: 4th-Generation Counterinsurgency
Dragon Days: Time for "Unconventional" Tactics
Terrorist Trail: Backtracking the Foreign Fighter
Militant Tricks: Battlefield Ruses of the Islamic Militant
Tactics of the Crescent Moon: Militant Muslim Combat Methods
The Tiger's Way: A U.S. Private's Best Chance for Survival
Phantom Soldier: The Enemy's Answer to U.S. Firepower
One More Bridge to Cross: Lowering the Cost of War
The Last Hundred Yards: The NCO's Contribution to Warfare

ADVANCED TACTICS IN AMERICA

ILLUSTRATED EDITION

H. JOHN POOLE

FOREWORD BY

MAJ.GEN. JOHN H. ADMIRE USMC (RET.)

POSTERITY
PRESS®

Published by Posterity Press
P.O. Box 5360, Emerald Isle, NC 28594
(www.posteritypress.org)

Cataloging-in-Publication Data
Poole, H. John, 1943-
Advanced Tactics in America
Includes bibliography and index.
1. Infantry drill and tactics.
2. Military art and science.
3. Military history.
I. Title. ISBN: 978-1-7354530-2-6 2022 355'.42
Library of Congress Control Number: 20222923200

Edited by Dr. Mary Beth Poole

First printing, United States of America, April 2023

For all U.S. infantrymen wishing to be as proficient as their forefathers.

Disclaimer: Those who find occasional references to the Creator unprofessional shouldn't bother to read this book. The author makes no apologies for such references, nor wants to hear how the book should have been better edited. Now that all mention of God has been removed from every U.S. classroom and courthouse, America may never acquire the cultural sophistication she has been seeking. The true measure of any society is in how it treats its most vulnerable members. The unborn certainly qualify, but so too do recent infantry recruits. Many U.S. leaders have trouble seeing how Christian principles can be followed and still conquer enemy evil. But, America hasn't decisively won a ground conflict since WWII. So, the Supreme Being is present in all things, and Godless warriors may not contribute much to the eventual peace.

CONTENTS

Illustrations ix
Tables xiii
Foreword xv
Introduction xvii
Acknowledgments xix

Part One: Ageless Knowledge
Chapter 1: *The Dangerous Illusion* 3
Chapter 2: *It's Harder to Fight a Thinking Opponent* 9
Chapter 3: *What's a Tactical Technique?* 13
Chapter 4: *The Need for More Small-Unit Maneuver* 25

Part Two: Past Conflicts
Chapter 5: *The French and Indian Wars* 33
Chapter 6: *U.S. War of Independence* 55
Chapter 7: U.S. *Civil War* 71
Chapter 8: *Western Indian Wars* 83
Chapter 9: *The Spanish-American War* 99
Chapter 10: *World War I* 115
Chapter 11: *World War II* 143
Chapter 12: *The Korean War* 161
Chapter 13: *The Vietnam War* 177
Chapter 14: *The War in Iraq* 189
Chapter 15: *The War in Afghanistan* 201

Part Three: Why America No Longer Wins Its Wars
Chapter 16: *Pentagon Has Failed to Evolve Tactically* 217
Chapter 17: *Shock and Awe Versus Deceptive Evasion* 223
Chapter 18: *Some New Hardware Discourages Maneuver* 229
Chapter 19: *Squad Tactics Still Vital to Modern War* 241

Part Four: Reshuffling the Priorities a Little
Chapter 20: *Washington's Perspective* 259
Chapter 21: *Battlefield Misconceptions* 265
Chapter 22: *The American Rifleman's Experience* 269

Notes
Source Notes 275
Endnotes 279
Glossary 307
Bibliography 313
About the Author 333
Name Index 335

ILLUSTRATIONS

Maps:

Chapter 9: *The Spanish-American War*
9:1 Outer Bastions of the Santiago Defense Belt 101

Chapter 10: *World War I*
10.1: Belleau Wood in 1918 126
10.2: Argonne Forest Took Its Toll in WWI 135

Chapter 12: *The Korean War*
12.1: Country South of the 38th Parallel 163

Chapter 13: *The Vietnam War*
13.1: CAP Locations at the Start of 1970 180

Chapter 15: *The War in Afghanistan*
15.1: Afghan Provinces 209

Figures:

Chapter 3: *What's a Tactical Technique?*
3.1: Sophisticated Daytime Assault Sequence 18-19
3.2: Not Every Team Must Get Through 21
3.3: Penetration Attempt Following a Diversion 23

Chapter 4: *The Need for More Small-Unit Maneuver*
4.1: On-Line GIs During Typical "Search and Destroy" Sweep 26

Chapter 5: *The French and Indian Wars*
5.1: French Commander's Scalp Saved at Lake George 35
5.2: Braddock's Men Are Ambushed 36
5.3: Rogers' Rangers Were Trained Like Indians 38
5.4: Optimal Viewing Conditions 50
5.5: Running Water Won't Obscure All Sign 51

Chapter 6: *U.S. War of Independence*
6.1: British Had Taken Terrible Losses at Bunker Hill 57
6.2: The U.S. Assault Line at Trenton 62

Chapter 7: *U.S. Civil War*
7.1: Fort Wagner Being Unsuccessfully Stormed 80

Chapter 9: *The Spanish-American War*
9.1: Battle of San Juan Hill 109
9.2: America's Fighters Were Determined but Mortal 113

Chapter 10: *World War I*
10.1: WWI Germans Depart from a Linear Defense 117
10.2: Assault Load of Early Stormtrooper 118
10.3: Just Looking Out of a Trench Was Risky 120
10.4: Unhappy Wheat Field Survivors Enter the Wood 128
10.5: The British Mark I Tank 130
10.6: Attack of the "Hell Fighters from Harlem" 138

Chapter 11: *World War II*
11.1: U.S. Units Weren't Prepared for Hedgerow Fighting 155
11.2: German Squad Strongpoint Matrix of WWII 157

Chapter 12: *The Korean War*
12.1: U.S. Lines Were Often Infiltrated in Korea 169

Chapter 13: *The Vietnam War*
13.1: Possible Vietnamese "Military Fortress" Segment 184
13.2: Helicopters Brought Casualties Out and Ammo In 187

Chapter 14: *The War in Iraq*
14.1: Quds Force Member in Lebanon 193
14.2: Lebanese Hezbollah Had Been Active in Iraq 199

Chapter 16: *Pentagon Has Failed to Evolve Tactically*
16.1: U.S. Rangers at Omaha Beach 221

Chapter 18: *Some New Hardware Discourages Maneuver*
18.1: NVGs Hinder as Much as Help 233
18.2: The Land Warrior 238

Chapter 19: *Squad Tactics Still Vital to Modern Combat*
19.1: Attack Objective from a Distance 245
19.2: Types of Gaps in Enemy Lines 245
19.3: Each of Parallel Lane Has a Pre-Confirmed Gap 246
19.4: A Single Buddy Team per Lane 246
19.5: Lead Men Move Toward Their Assigned Route 247
19.6: One Team Follows a Watershed Ditch 247
19.7: Adjacent Team Enjoys Best Infiltration Route 248
19.8: Third Must Mask Noise of Moving Through Brush 248
19.9: Clearings Take Camouflage and Very Slow Movement 249
19.10: Foe Becomes Visible Against Lighter Background 249
19.11: Stalker Keeps Object Between Himself and Quarry 250
19.12: Sapper Team Approaches Jump Off Point 250
19.13: Lead Man Visualizes Move While Studying Gap 251
19.14: Forward Member Departs After Sentry Distracted 251
19.15: Backup Man Draws "Bead" on Sentry's Head 252
19.16: Penetration Occurs During a Diversion 252
19.17: If That Sentry Gets Nervous 253

TABLES

Chapter 3: *What's a Tactical Technique?*
3.1: Penetration Sequence for Barbed-Wire-Free Objective 22

Chapter 6: *U.S. War of Independence*
6.1: The "Maneuver Warfare" Difference 66

Foreword

John Poole's intellectual and common-sense thoughts on war are insightful and thought-provoking. In *Advanced Tactics in America,* he describes and compares regular and irregular as well as historical and modern concepts of war. His writings motivate us to think, intuitively and innovatively, about the simplicities and complexities of war. War's brutal and ruthless, but it's often necessary to protect our cherished freedoms.

Karl von Clausewitz, a renowned theorist on war, wrote in his classic treatise, *On War,* that tactics is the employment of forces to win battles and strategy is the employment of battles to win wars. While this is a true and simple explanation of strategy and tactics, both are exceedingly complex, too. While he wrote of warfare during the Napoleonic Era, his strategic and tactical theories are timeless and remain essential to war today.

But war is about more than strategy and tactics. Accordingly, Clausewitz theorizes that war is simply an extension of politics by other means. Therefore, success in war depends upon a more holistic approach that includes statesmen, soldiers, and citizens. We can never simply send our military and soldiers to war. We, our statesmen and citizens, must support and send our Nation to war. We must be all in to win. All in wins.

Today, as in yesterdays, the basic challenges and dilemmas of war are to create relatively precise strategies and systems, rather precise tactics and technologies, for relatively imprecise future contingencies. War and its weapons, as well as strategies and tactics, inevitably evolve and rarely, if ever, remain constant. Nations and militaries who progress often succeed; those who accept the status quo or regress often fail.

Julius Frontinus, Chief Military Engineer to Roman Emperors, circa 70 AD, wrote: "I will ignore all ideas for new works and engines of war, the invention of which has reached its limits and for whose improvement I see no future hope." Poole has an opposing philosophy: future success depends on blending consistent innovative technologies and tactics with the irrefutable traditional values of the individual combatant and small units.

While Poole focuses primarily on strategies and tactics, he understands the vast issues and influences on war. Militaries are a combination of ground, air, naval, and space force options. Further, nations have nuclear, conventional, and special warfare capabilities. The challenge becomes designing and deciding on force compositions and capabilities as well as their sizes and funding. These perplexities defy simplicity.

Yet, success in war is about more than state diplomacy, military prowess, or citizen support. It's also about industrial, economic, and cultural strengths of nations and peoples. As WWII was ending, for example, Hitler declared it was not the U.S. military that defeated Germany, but the American worker. We simply outproduced Germany with our industrially and culturally inspired blue collar work ethic. Frontline Soldiers and Homefront Workers are both essential.

Similarly, President Reagan later denounced Soviet socialism and communism, with their rigid centralized control, as destructive to economic incentives for innovation as well as their denial of individual freedoms for self-achievement. The dual principles of democracy and free enterprise ultimately defeated the twin evils of socialism and communism. Self-reliant nations and peoples armed with freedom and the right to self-expression often prevail. Multiple force composition and capability options, combined with diplomatic, industrial, economic and cultural powers, cause the complexities of war to become exponential, especially in an unpredictable future. War's critical decision-making process has crucial consequences for the fate of nations and peoples. The future might be unknown, but we must nonetheless plan and prepare for it to protect, preserve, and perpetuate peace.

Nations recruit, equip, train, resource, and deploy militaries for three basic reasons: to fight at the right time, at the right place, for the right reason. All are important, however, fighting for the right reason, a justifiable and honorable cause and peace, is crucial. In many respects, there is no glory in war, but there is often glory in the true and just peace it can bestow upon its belligerents, both winners and losers. We must fight as hard for peace as we do for war.

It's statesmen in pre/post-war diplomacy who must initiate plans and policies to prevent an unjust war as well as fight a just one. It's soldiers who must endure the sacrifices and miseries of war. It's citizens who must understand the burdens they impose on those soldiers they send to fight on their behalf. A nation and its peoples have a moral obligation to respect the profound impact war has on those who plan, fight, and support it.

Within this context, Poole acknowledges the crucial significance of statesmen, soldiers, citizens, democracy, industrial, and economic factors in war. He devotes considerable attention and respect, however, toward those individual soldiers and small units who engage in the terrorizing and traumatizing aspects of war. It's those soldiers in the frontline trenches and deadly confrontations who Poole most respects . . . and writes to protect.

Poole provides historical and contemporary examples and illustrations for training and employing today's individual combatants and small units. In numerous scenarios, with compelling details for actions and counteractions and reactions, he presents tactical options. In the process, without necessarily or indirectly stating them, he cautions, rather emphatically, that we must avoid and never succumb to four basic myths of war.

First, the next war will be similar to the last war, but wars often vary in their tactics, technology, and tragedy. Second, the war will be brief, but wars often become lengthy and exceed expectations. Third, the war will be relatively inexpensive, but wars often become terribly costly and expensive in lives and resources. Fourth, the war will be controllable, but wars often escalate and become uncontrollable.

As a veteran, writer, and fighter in war, Poole understands the challenges and dilemmas of war, especially from the perspective of the individual combatant and small units with whom he is so familiar. He understands the heart of war, its human emotions, its raw and painful cruelty, and its soulful impact on those who fight it. Poole is a true advocate for the individual combatant and small unit in war and he champions their initiative and agility.

In summary, efficiency and effectiveness in war and peace involve diverse, multifaceted, and complex individual entities and national interests, which range from huge government bureaucracies and corporate conglomerates to individual combatants and small units. The challenge is to integrate the many to interact with one another in the interests of peace. These are the challenges and complexities and contexts of war within which Poole writes.

Ultimately, it's well-trained, well-equipped, and well-led individual combatants and small units who fight battles that win wars. This is Poole's unequivocal mantra in this book and his writings as a staunch advocate for warriors trained and encouraged to exercise initiative and agility on the battlefield to become a powerful force multiplier in war. Poole hails our warriors who fight with courage and conviction for our freedoms and peace.

MAJ.GEN. JOHN H. ADMIRE USMC (RET.)
FORMER HEAD OF 1ST MARINE DIVISION

Introduction

Tactics are combat maneuvers. They can be performed by any infantry echelon down to, and including, an individual rifleman. But, "advanced tactics" normally require a composite mixture of varied moves from elements of squad size or smaller. Such a varied effort is seldom encouraged by the United States (U.S.) military because it flies in the face of standardization and top-down control.

Surprise and firepower are interchangeable in war.[1] The more secretive the attack, the less shooting will be required to complete it. Weapon sophistication also has an effect on method. The latest models make killing possible from a distance. That's when "boots on the ground" seem unnecessary, and their short-range proficiency extraneous.

Well-armed units will often employ linear formations to maximize their firepower. Any tiny element to the front of such a formation partially masks its fires. So, most patrols and outposts are pulled in before any enemy contact. Through too little challenge, they eventually lose their self-sufficiency.

The purpose of such a discussion is something the average U.S. veteran may find hard to accept. While the above factors were shaping America's standard operating procedures (SOPs), less-well-endowed nations were focusing on ways to defeat the latest technologies. And, Yank infantry forces became among the best-equipped, and least agile in the world. Better are the small unit capabilities of traditional adversaries—Germany, Japan, Russia, China, and other Asian Communist nations. Until something is done about this frontline deficiency, America risks losing World War Three (WWIII).

Cpl. Joe Friday of Paducah would want to know how such a mismatch could be possible. He is proud of his service in Iraq and Afghanistan. His father and grandfather had spoken equally fondly of their contributions to Vietnam, Korea, and World War Two

(WWII). All had arrived at good short-range stratagems on the spur of the moment. But, tired soldiers will fight as trained despite any extenuating circumstance (like, microterrain available on offense, or claymore mines in the way on defense). So, the Stateside military never hears of enough other examples of the same brilliant idea to justify an SOP change.

As anomalies to the normal regimen of war, most lessons learned aren't well enough corroborated to be included in the after-action reports. Through over-editing, others escape the chronicles and manuals. In their place appear simplistic procedures that noncommissioned officers (NCOs) too often use as discipline-building drills. Even when memorized, such procedures can't generate enough surprise to be of much value in combat. Under ideal circumstances, they result in excess casualties. Yet, all U.S. infantry training still revolves around them. And each new generation of GIs has to "learn the hard way" what will actually work in a close Eastern encounter.

While previous Posterity Press studies have mostly focused on enemy skills, this book will search for advanced tactical techniques from U.S. history. All will be logical and help when the latest gadgets wet, shot, or otherwise broken. Most light-infantry methods are timeless—just as viable now as when originally dreamed up. When the appropriate method comes from government issue (GI) forefathers, the current generation may more easily accept it as legitimate.

Through too little attention to the situational details of war, most U.S. infantry outfits have become less tactically adept than Eastern counterparts (whether regular army or local militia). Top-down Western societies too heavily focus on the big picture. And, excellence in battlefield maneuver doesn't necessarily increase with the passing years. So, be prepared for a shock when "advanced" small-unit moves were last practiced by many GIs.

LT.COL. H. JOHN POOLE USMC (RET.)
CAREER INFANTRYMAN AND
FORMER FLEET MARINE FORCE GY.SGT.

Acknowledgments

Once again, God must be given the credit for any useful insights from this latest research. Tiny-element techniques will continue to be the key to decisively winning with minimal casualties in war. According to Pope John Paul II, the most moral form of combat causes the least damage to, and possibly saves the life of, the opponent.[1] That requires more "moving" than "shooting." This book will "communicate" how.

PART ONE

AGELESS KNOWLEDGE

"STRATEGY WITHOUT TACTICS IS THE SLOWEST ROUTE TO VICTORY."
– SUN TZU

(Source: Sun Tzu, "The Art of War.")

1 The Dangerous Illusion

- Has Pentagon fulfilled its most recent responsibilities?
- In which areas might it improve a little?

The indestructible American soldier.

(Source: U.S. Air Force (www.af.mil), likely drawn by a paid member of the U.S. Armed Forces, this public domain image has image designator, "AFG-060725-004.jpg.")

What's Now Missing from the Equation

All wars are waged through fire and maneuver. But, with U.S. forces, the amount of the latter has been steadily decreasing. Only in open desert can motorized infantrymen achieve sufficient agility. Everywhere else, too much broken terrain stands in their way.

U.S. Ground Combatants Are the Best on Earth

U.S. infantrymen are repeatedly told they're the world's finest.

They're the best equipped for mechanized warfare, and the best at following orders. Most have unlimited growth potential. But, claiming they're the best at all things is a bit of a stretch. American riflemen aren't permitted the initiative and agility of an Asian Communist soldier.

> It is people, not things, that are decisive in war.[1]
> — Mao Tse-tung

> [A]n attached observer has described Mao's soldiers as "constantly studying their mistakes, and improvising methods to . . . offset modernized [Western] equipment."[2]
> — Wukovits, *American Commando*
> and "Voice of America," September 2010

> [I]n . . . regular warfare, . . . [armed struggle] is fully capable of . . . getting the better of a modern . . . U.S. Army. . . . This is a development of the . . . military art . . . which is to rely chiefly on [the] man, . . . to defeat an enemy . . . [having] up-to-date weapons and equipment.[3]
> — Gen. Vo Nguyen Giap
> *The Military Art of People's War*

Yet, GI superiority is still touted as a morale builder aboard most U.S. military bases. American grunts are issued the latest electronic gadgets. And "hopeful exaggeration" still makes its way up the chain of command. At one WWII battle (possibly Buna on New Guinea), Gen. MacArthur declared victory three days before it actually happened. Since then, occasionally stretching the truth has become an acceptable part of U.S. military decorum. U.S. forces have become so mobile that dismounted infantrymen are no longer necessary. Budget writers seldom disclose any training shortfall. Commanding officers consider enemy strengths an inappropriate entry on any after-action report. And supply specialists replace "aging" maneuvers with technological crutches. That's when this chorus of over-optimism can adversely affect the U.S. ground pounder's chances of surviving.

If GIs really were the world's best at all things, they would have no reason to improve. There would be no competition between units. No battle chronicle would provide additional insight. In short, there would be no spirit of improvement.

Good Intentions Aren't Enough

Within the U.S. military, unit training normally takes precedence over individual instruction. But, when lowly grunts are exposed to advanced combat moves, the bar is set too high. To enter a house window, not all will have the dexterity to do so as demonstrated. Nor will they get to develop their own version. At that point, their unit only appears to be ready.

In war, every infantry unit largely depends on its weakest link. Without enough all-hands cooperation, it can fail at the simplest maneuver. That's why each U.S. rifleman should be allowed his own portfolio of personalized, yet still combinable, techniques. During a fire and movement exercise, he must flop down. How best to do so varies between body styles.

The Equipment Issue

GIs do enjoy the most advanced gear in the world. It's issued so often there's little time to train on anything else. But, no new device can replace dodging bullets. Wars cannot be won from standoff distance, and most future engagements will occur at ranges too short for the latest shoulder-fired rocket. The shortcomings of new Western gear would take to long to itemize. Suffice it to say, ground warfare hasn't changed much since the machinegun. And all Eastern powers have ways to counter subsequent inventions. Simply moving below ground will work. For example, very little of the Allies' heavy ordnance on Iwo Jima, Peleliu, or Okinawa made any difference at all.[4] Then, a fully supported U.S. phalanx would be kept from entering North Korea in 1953,[5] and be again stymied in Vietnam and Afghanistan.

In fact, there may be a direct correlation between new equipment and tactical obsolescence. If too much emphasis were placed on mechanization, wouldn't all infantry outfits lose their mastery of the microterrain?

What the Manufacturers Have Failed to Mention

New inventions still get damaged in combat. On Tarawa, there was no radio communication for three days.[6] On Okinawa, too few

81mm mortar rounds were on hand at first.[7] In Vietnam, M-16's were still jamming after one shot as late as August 1968.[8] So, no matter how sophisticated, any new equipment requires a physical backup. An old infantry procedure can often provide it. The GPS (Global Positions System), for example, is virtually useless in heavy jungle. Expert land navigation is its only substitute.

Most new ordnance is to kill opposition soldiers from farther away. With each new model comes less dependency on close combat. That it is bulky, noisy, and temperamental makes its wearer less stealthy. It in no way obscures him from enemy detection. Small-unit maneuver excellence is measured by its degree of surprise.

But, U.S. squads, fireteams, buddy teams, and individual riflemen have never been very tactically sophisticated, because they haven't been required to be. Any member of an American on-line sweep who stops to deal with enemy opposition will lessen that sweep's momentum. And any tiny element out in front of such a formation will not only mask its fires, but risk taking friendly fire. So, U.S. infantry squads don't get enough practice at working alone to become self-sufficient. They would if they could. In fact, they could do as well—while severely outnumbered—as any foreign guerrilla group.

> The emphasis on modern-day firepower has been so great that well-meaning commanders have overstressed safety in their maneuvers.[9]
>
> — *Army Magazine,* August 2020

Origins of This Tactical Oversight

Once too many machineguns to suppress are acknowledged, an on-line sweep or row of sentries becomes too risky to attempt. The British and French were slow to realize this in World War I (WWI), and the U.S. Pentagon has yet to. Those linear formations are best replaced by squad strongpoint belts on defense, and squad-sized probes on offense. Or, for Unconventional Warfare (UW), there can be ambushes on defense, and infiltrations on offense.

Why doesn't long-range small-arms fire ever bother U.S. invasion planners? Their bombardment was thought heavy enough to overcome any form of resistance—be it enemy thinking, bullet streams, barriers, tactics, or terrain choice. If foe small-unit maneuvers are no longer a concern, there's no reason to try to match their sophistication.

Cultural Pressures

America's soldiers and Marines are as good as they come at every rank. Yet, most enlisted personnel consider their officers poorly informed, and most officers have yet to discover enlisted wisdom. The reason is top-down hierarchy. Those in charge are kept from frontline problems by subordinates protecting their own careers. The longer the chain of command, the bigger the communication gap between general and actual fighter.

The Big-Picture Lacks Essential Detail

Few Americans yet realize the different way that most Easterners solve problems. Nor are they aware of the Eastern belief that opposites can both be true, as in the *yin* and the *yang*. Such a dichotomy is most easily explained through big and little pictures. While the top-down Western thinker focuses on the general scenario, his bottom-up Eastern counterpart masters composite parts. Their solutions then widely differ. While the overall situation may appear favorable, a few of its most vital aspects have gone astray. And thus occurs another kind of illusion.

3rd Generation Warfare (3GW) is where the foe's frontline strengths are bypassed to more easily get at his command and logistics hubs. That's why 3GW's parameters are often the opposite of those from 2GW (killing as many enemy soldiers as possible). And not all U.S. tacticians associate the 3GW kind with Asia. Some believe them to have come from WWI Germany. Either way, bottom-up Eastern cultures have less trouble adopting the more moral of the two—3GW.

National Security Is at Stake

America's frontline fighters have not been lied to exactly, just shown parts of the truth. But to have the best chance of succeeding, they must come to know more of it. A briefing on the big picture won't be enough. Their world will soon boil down to a barely visible automatic-weapon muzzle 30 yards away. No modern-day bunker-busting missile will arm in time, nor will a grenade launcher be accurate enough. Ultimately required will be a hand grenade through its tiny aperture. Only if a couple of "liberty risks" can figure out how to cross those 30 fire-swept yards will that enemy weapon be silenced. Ground wars are many such episodes in series. So, small-unit maneuverability becomes strategically essential. The only way to stop the beachfront slaughter at Peleliu, had been a pair of Leathernecks crawling up on "the Point's" 30mm automatic cannon.[10]

2 It's Harder to Fight a Thinking Opponent

- Do all armies try as hard as they can to win?
- Why might a few disregard the foe's frontline stratagems?

Many Muslims enjoy a heritage of small-unit action.

(Source: DA PAM 550-25 [December, 1988], chap. 4.)

The Western Presumption

Top-down Western societies expect their senior officers to win the battles. If the foe is at all fooled by the initial attack, the U.S. commander's scheme of maneuver can initially work. But, after facing the right small-unit countermeasures, it may later fall apart.

What's by Definition Two-Sided

Warfare is "armed conflict between two massed enemies, armies, or the like."[1] The winning side has normally paid close attention to the other's frontline agenda. Otherwise, what appears to be a "slam-dunk" won't be.

Where There's Smoke

Over the years, low-ranking American riflemen have been allowed very little constructive thinking ("master-minding" as it was called in WWII). Their more organizationally savvy superiors took care of that. Many recruits were thought to be too free-spirited to make good soldiers. Only after being thoroughly disciplined might they successfully join an infantry team.

To overcome this supposed deficiency in the American psyche, several things were arranged. The enlistment oath required strict obedience to orders,[2] and most initiative at boot camp was punished. Without any further discussion, all instructions were to be instantly obeyed. And, during the follow-on infantry training, every battlefield eventuality was supposedly explained.

But, close combat is not well structured, nor can it be made that way. It will always be fast moving with unsuspected challenges. At some point, that young American will encounter a threat for which he has received no instruction—by any leader or school. What does he do then? Hopefully, having to think for himself won't overly slow his reactions. And, under-preparing that Pride of Poughkeepsie may have other ramifications.

If the enlisted personnel of the greatest democracy aren't using their heads much, then those of a foe dictator aren't thought to be doing so either. So, U.S. invasion planners make no provision for the innovative small-unit maneuvers of an opportunistic bad man.

More Than One Warfare Style Is Valid

Just as different ways of looking at things exist in every marriage, so are opposing perspectives legitimate on the battlefield. Just for starters, there is the "top-down" Western viewpoint, and the "bottom-up" Eastern viewpoint. Both have produced valid warfighting methodologies.

Overwhelming force through 2GW is the West's favorite way of operating, but the Eastern-rooted 3GW is equally powerful. Otherwise, the U.S. Marines wouldn't have made Maneuver Warfare (MW) their new doctrine in the late 1980's. One way tries to win through killing the foe's frontline soldiers, and the other through disrupting his rear-area centers of gravity.

But, fully practicing 3GW takes self-sufficiency at the lowest echelon of every infantry outfit, and that flies in the face of two U.S. traditions—method standardization and centralized control. Too much delegated authority and semi-autonomous small-unit activity might jeopardize the parent unit's mission. And, because 2GW demands the absolute loyalty of every participant, any move toward a full 3GW capability is tantamount to sedition.

Though 3GW is more moral in its application, Americans see no problem with its partial acceptance. U.S. citizens think their army the most advanced at all things. That makes all Eastern counterparts (including the Communist ones) "underdeveloped" at the same things. But, high-tech 2GW in America is not as humanistic as it appears on a distant TV monitor. And the East's most productive procedures predate Communism by thousands of years. Though largely democratic, those procedures are still allowed in many highly authoritarian armies.

> [T]he reason why the Red Army can sustain itself . . . in spite of such a poor standard of material life and such incessant engagements, is its practice of democracy. . . . [O]fficers and men receive equal treatment; soldiers enjoy freedom of assembly and speech; . . . and the books are open to the inspection of all.[3]
>
> — Mao Tse-tung

> In *Kiem thao* sessions, the [North Vietnamese] soldiers offered judgments of their comrades and listened to evaluations of their own performances. The meetings sometimes featured discussions of tactics from the unit's recent engagements or *suggestions . . . sent from the army command* [Americans would call this policy]. *Kiem thao* sessions could become extremely heated and emotional. For some soldiers the sessions were especially traumatic, as they heard their weaknesses and failings denounced publicly and then had to respond to those charges. [Italics added].[4]
>
> — *Vietnam Experience: A Contagion of War*
> by Maitland & McInerney

Might the American Way of War Be Improved a Little?

Neither thinking small nor collective effort are integral parts of Western culture. Roadside sniping had thwarted a British column at Concord in April 1775, but Americans don't normally fight that way. There was no cannon fusillade at that first military engagement of the War of Independence. All the U.S. Minutemen needed to discourage British regulars were a few well-directed musket shots.[5]

So, the answer to everything warlike need not be foreign. Those frontier militias had learned from the Native Americans how to handle a concentrated thrust. Though counter-cultural, it had worked. No culture is perfect, particularly the younger ones. All have bad habits. So, any military organization that allows its civilian culture to shape its wartime capabilities is just asking for trouble.

Must a nation's army have the same blind spots as its parent society? Or should its ground combatants be allowed to compensate for any communal weakness? The U.S. military hasn't decisively won a ground war since WWII. Did giving every soldier and Marine the same freedoms as the average citizen keep the Pentagon from developing enough cohesion and small-unit excellence to achieve final victory?

3 What's a Tactical Technique?

- What makes a maneuver tactically sophisticated?
- How does generating more surprise help its user?

What up to 14 people can do without any supervision.

(Sources: FM 22-100 [1983], p. 84; FM 7-70 [1986], p. D-24.)

What Constitutes Technique in Wartime

A "technique" is a situationally tailored way of accomplishing some military task. A "tactical technique" normally involves the offensive maneuver of a squad or subordinate element. It is a group effort with no need for a leader. With enough practice, that effort can be through muscle memory alone. That catches the enemy defender off guard.

In the Western World, almost every battlefield action is pre-planned or directed from above. But, because few instructions to lower-echelons are comprehensive, a lot occurs on the front lines instinctively. There, each set of circumstances must be approached in one of two ways—through a well-focused unit-wide technique, or somewhat appropriate organizational method. The first usually works better than the second.

With enough imagination, advanced tactical technique can be acquired with "bottom-up" training through consensus opinions and field experimentation. However, the Asian Communists enjoy another source. They learn through fighting. Successful fireteam moves are added to the squad portfolio.

> Reading books is learning, but application is also learning and the more important form of learning. To learn warfare through warfare—this is our chief method.[1]
>
> — Mao Tse-tung

This approach works for the Red Chinese because every unit, however small, must be ready to quickly shift between Positional (2GW), Mobile (3GW), and Guerrilla Warfare (GW).[2] Within the latter lies the building block for good squad technique.

> The strategy of guerrilla war is to pit one man against ten, but the tactics are to pit ten against one.[3]
>
> — Maxim from Mao Tse-tung

Because GW is mostly conducted by tiny, outnumbered elements, it automatically generates advanced (surprise-oriented) techniques. Such techniques are then available for other tasks. It takes highly secretive small units to sidestep enemy concentrations in 3GW. And winning several tiny skirmishes in a row can develop as much momentum as a regimental thrust. The mosquito becomes more of a threat when it swarms.

World-Class Squads Have Alternative Maneuvers

How to acquire tiny component leeway and a controllable whole at the same time has largely escaped U.S. leaders. Most want every aspect of war to be standardized, so that their instructions will be more completely followed. But, that much assurance produces deadly predictability in combat.

A football team couldn't go far down the field with the handful of simplistic squad maneuvers in a U.S. infantry manual. At a minimum, variations that better mesh with unique circumstances would be necessary. For example, a daytime short-range infiltration might work better in tall grass than a stand-up charge.

If nature were allowed to shape when to fight, all armies would become more small-unit proficient with time. But, top-down military bureaucracies resist every change, however useful. When former SNCO (Staff Noncommissioned Officer) and future MoH (Medal of Honor) winner Wesley L. Fox took over his well-seasoned rifle company in Vietnam in 1968, he discovered a complete absence of tactical technique. For that unit, bureaucratic pressures had led to the loss of almost every squad, fireteam, buddy team, and rifleman survival trick since WWI.

> Using football as an analogy, my players knew how to play the game, but had no team plays, much less any rehearsal of these plays. . . .
>
> " . . . My purpose is not to teach you anything but to pull our collective knowledge together, to get our purpose, our plays, our timing down to a razor edge."
>
> " . . . We will spend two hours each day here talking and walking ourselves through some offensive tactics that I feel are needed."[4]
>
> — 1st Lt. Wesley Fox
> upon assuming command of A/1/9 in Vietnam

Desperate Circumstances Had Produced Many Tricks

Marines who dashed across the isthmus on Iwo Jima within full view of Mount Suribachi and the southern plateau didn't do so upright. They were at least crouching.[5] And, they mostly moved between shell holes (without ever lingering for long). With elevated enemy observers watching their every move, those young Leathernecks had been trying to avoid a predictable choice.

This original dash had provided two vital lessons. The first was to bypass enemy resistance when possible. The second was for each rifleman to defeat the first bullet through an unexpected "juke." Hurriedly crossing the isthmus had produced a series of nonstandard moves that would pay huge dividends later. Atop Iwo's northern plateau, the terrain was so broken that only fireteams could maneuver. That's where those new and often individualized moves would make the most difference.

> They [the Marine riflemen] crept and crawled, dodged and ducked, slithered and staggered—but they moved forward.[6]
>
> — Bill Rose, a Marine combat correspondent

After such a freewheeling offense on the first day at Iwo, the SNCO "chain of command" saw their chance to help. During any subsequent approach march, all lower ranks moved a little quicker, ducked more, and "hit the deck" before any foe could acquire a three-second sight picture. Movie footage exists of a small Marine column zigzagging—at a trot—between tiny promontories in what was probably the prelude to an attack. No 2GW assault procedure allows its participants to get all the way down. Yet, Marines had partially crawled during a patrol-like seizure of Suribachi's crest, and sooner or later would have crawled on line during the final stages of any contested assault.

Staying Healthy Takes Bullet Avoidance

On an active battlefield, all above-ground space can be filled with whizzing metal. That's why the motto of the 6th Division on Okinawa was "stay off the skyline." There's only one way to avoid all that metal—utilizing the microterrain. That takes staying behind rises, ledges, mounds, rocks, stumps, fallen trees, etc., or inside gullies, ditches, holes, etc. Just hiding in the vegetation won't get the job done.

When a U.S. squad member is told to shift position in a hurry, all that microterrain is no longer available to him. Nor can any upright movement skills save him for long. Until American maneuvers regularly allow a little crawling, too many young people are going to get hurt.

The U.S. Army found so many units getting pinned down by automatic weapons during WWII and early Korea that its infantry conferences dictated a "bounding overwatch."[7] That method obligated separate elements to keep going against any amount of incoming fire. And, the least acceptable way of doing so was for a whole contingent to slither forward during its turn to move. That allowed too little overall momentum. So, people got shot during a bounding overwatch. And the U.S. Army considered their loss necessary. Needless to say, waiting for nightfall would have also limited casualties. Any speed faster than the enemy expects will achieve momentum.

Some Techniques Must Be Done by the Numbers

As with the "blind 'em, burn 'em, and blast 'em" procedure of early WWII in the Pacific, most 2GW techniques have to be precisely executed. That's why instantaneous obedience to orders was so important at boot camp. If the steps of a 2GW technique aren't accomplished in close succession, the enemy may have time to regroup.

There is also inherent risk in the full-force of any 2GW method. Some create dangerous shrapnel and backblast. If participants don't get down at the right times, they may be injured and the method fail. The daytime assault in Figure 3.1 is more advanced than any dictated by the U.S. military. Developed at the Marine Corps' East Coast Platoon Sergeants School in the late 1980's, it adds extra steps to the 2GW routine—with more surprise resulting. But, even here, the SMAW (modern-day bazooka) man must duck after taking his bunker shot or lose the top of this head to the bangalore. Nor can anyone wander into his backblast area.

3GW Techniques Allow More Initiative

For each infantry squad, 3GW procedures involve more subsidiary content, movement options, and internal decision-making. In the short-range-infiltration of Figure 3.2, each two-man team confronts different challenges in its lane, and uses separate diversions to sneak between enemy holes.

Among 3GW's approach march skills would be: (1) quieting all gear; (2) camouflaging to match surroundings; (3) crouching when not crawling; (4) so slowly moving across open ground to be imperceptible in low light; (5) staying in defilade most of the time; (6) following the ripples of the earth; (7) staying off the topographical or military crests of tiny hillocks and mounds; (8) avoiding false horizons (lighter backgrounds); (9) shadow walking; (10) keeping a bush or mound between oneself and the sentry; (11) avoiding any water

Figure 3.1: Sophisticated Daytime Assault Sequence
(Sources: FM 22-100 [1983], pp. 22, 30; FM 24-1 [1976], p. 3-9; FM 5-103 [1985], p. 4-7; FM 7-11B3 [1976], p. 2-VII-C-4.4.)

reflection; (12) not making an unnatural shadow; (13) not causing unnatural motion in tall grass; (14) remaining stationary beneath a flare; (15) crossing a road or trail at a low spot, between curves, or in deep shade; (16) moving quickly between shadows in moonlight;

Figure 3.1: Sophisticated Daytime Assault Sequence (Continued)
(Sources: FM 23-30 [1988], p. 4-6; TC 90-1 [1986], cover; FM 5-103 [1985], pp. 4-4, 4-6.)

(17) following a low row of vegetation to cross a farmer's field; (18) staying downwind of the quarry; (19) waiting for a background noise to cross dry leaves or swampy ground; and (20) trimming bottom twigs off obstructing bushes.

Diversions for more easily slipping between sentries might come from: (1) a cloud covering the moon; (2) a flare behind the infiltrator; (3) a sentry smoking, eating, dozing, being relieved, being inspected, or being bothered by insects; (4) the background noise of a helicopter, tracked vehicle, artillery barrage, or weather; (5) the sound from another lane; or (6) an intentional distraction from one's own partner.

Among the possible gaps in the enemy line would be: (1) a watershed ditch; (2) uncleared vegetation; (3) damp ground; (4) a tree's shadow; or (5) a fallen log.

Unsuspected dangers might include: (1) thermal imaging from a foe contingent too small to normally have it; (2) a yet-to-be spotted forward listening post; (3) a roving sentinel who has discovered the infiltrator's trail; or (4) a hidden defense line.

Clearly, the 3GW warrior requires more training and imagination than his 2GW counterpart. While instant obedience to orders is still important, so too is identifying which orders lack enough detail. The combatant then uses every known trick to follow the intent of his orders within the allotted time. Any necessary action for which he has no prior knowledge, he visualizes. A short-range infiltration is his favorite form of assault, because not all teams must get through. Those who don't feel comfortable with an entry can cover the escape of others.

Where there is no barbed wire, the 3GW option is normally best. Through "recon pull" and less risk taking, it endangers fewer people. A stand-up assault against enemy strength may have worked in the days of the unrifled musket, but it won't against hidden machineguns. (See Table 3.1 and Figure 3.2.)

Techniques Must Be Flexible

As in football, very few assault techniques can be run exactly as rehearsed. If someone or something is in the way, participants must get around the obstruction. Rehearsing in different types of terrain should give them enough muscle memory to execute the original plan. But, as with football plays, there are tactical tech-

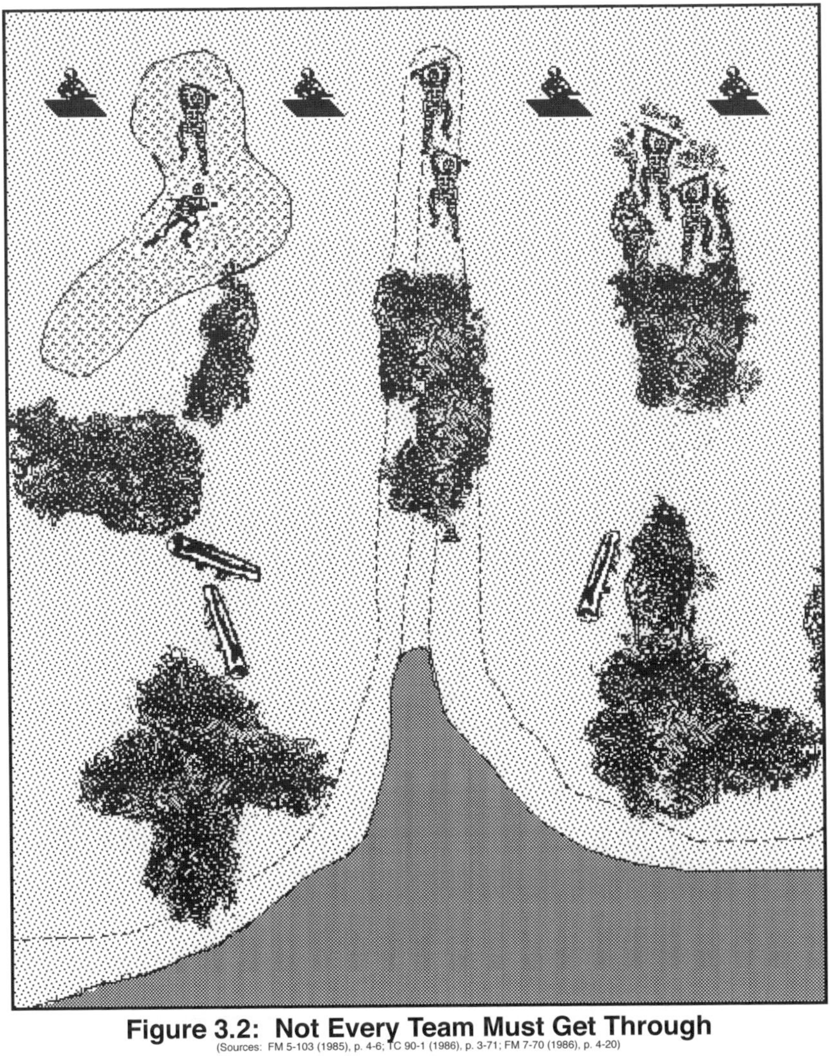

Figure 3.2: Not Every Team Must Get Through
(Sources: FM 5-103 (1985), p. 4-6; TC 90-1 (1986), p. 3-71; FM 7-70 (1986), p. 4-20)

PRIOR TO THE PENETRATION ATTEMPT

THROUGH BINOCULARS FROM A TREE, THE TWO U.S. INFILTRATORS INSPECT THEIR PRE-RECONNOITERED LANE AND ITS CHOICE OF GAP

THE TWO "SAPPERS" STRIP OFF ALL EQUIPMENT EXCEPT FOR EXPLOSIVE, TIMER, CARBINE, AND BACK-SLUNG WATER

THEY CAMOUFLAGE THEMSELVES TO LOOK LIKE WHATEVER MOSTLY FILLS THEIR LANE—BUSHES, GRASS, OR ROCKS

DURING THE APPROACH

AT DUSK, THE PAIR MOVE UP THEIR LANE WELL AWARE OF A POSSIBLE OUTPOST WITH ROVING SENTINELS

THEY HAVE CHOSEN A FULLY MOONLIT NIGHT SO THE ENEMY'S NIGHT VISION DEVICES WILL BE USELESS

IF THE DEFENSIVE FORCE IS BIG ENOUGH TO HAVE ITS OWN THERMAL IMAGING, THE SAPPERS WAIT FOR A RAINY NIGHT

UPON SPOTTING THE SENTRY IN THEIR LANE, THEY STALK HIM—KEEPING SOME MOUND OR VEGETATION WITHIN THE INTERVENING SPACE

TO CROSS OPEN GROUND, THEY GO SO SLOWLY THAT THEIR MOVEMENTS WILL BE INDISCERNIBLE THROUGH THE AVAILABLE LIGHT

THEY GET WITHIN 20-30 FEET (GRENADE RANGE) OF THE ENEMY DEFENDER IN THEIR LANE BY 2:00 A.M.

BYPASSING THAT SENTRY

FROM 2:00 TO 3:30 A.M., THEY STUDY THE GROUND THEY MUST CROSS TO GET PAST THE SENTRY, LOCATE THE ADJACENT WATCH STANDER, VISUALIZE EVERY PART OF THEIR PASSAGE, AND THEN WAIT FOR A DIVERSION

WHEN THAT DIVERSION COMES, ONE SAPPER TAKES A BEAD ON THE SENTRY'S HEAD WITH HIS RIFLE, AND THE OTHER ADVANCES

AFTER PASSING THE SENTRY, THE FIRST SAPPER TAKES A BEAD ON THE BACK OF THE SENTRY'S HEAD, AND THE SECOND ADVANCES

GETTING BACK OUT

AFTER COMPETING THE MISSION, BOTH REVERSE THE ENTRY PROCEDURE.

Table 3.1: Penetration Sequence for Barbed-Wire-Free Objective

Figure 3.3: Penetration Attempt Following a Diversion
(Sources: FMFM1-3B [1981], p. 4-9; FM 5-103 [1985], p. 4-4)

niques not well enough designed. The U.S. military has long filled its manuals with overly simplistic small-unit maneuvers. Their supposed purpose was ease of adaption. But, those maneuvers weren't published as guidelines. And, heavily loaded troops in jungle heat with only power naps for sleep are not that opportunistic. When in deep *kimchi*, they'll do what they can remember and hope for million dollar wound.

To add insult to injury, people with drill sergeant mentalities have been using those simplistic maneuvers as battledrills to enhance discipline. That creates bad habits within the lowest ranks. When extremely tired, a junior grunt might try one of those fully predictable maneuvers against a world-class light infantryman from Asia.

The U.S. enlistee's early military life revolves around doing precisely as told, from strictly obeying his 11 General Orders to memorizing the immediate-action drills in his squad manual. Too greatly relying on the counter-ambush drill could have disastrous results. That drill calls for an upright assault against any ambusher less than 50 yards away. That's too far to cross standing up. If it contains heavy brush, tired Americans might mistake screening for cover. Then, a single automatic weapon behind a low berm could knock down every GI. And, the ultimate culprit would be a method too hastily designed by staff officers, over-edited in publication, and then used to build aggressive spirit. The automaton has a limited life span, whereas a thinking rifleman may help to raise the final flag of victory.

4 The Need for More Small-Unit Maneuver

- When does the tiny mosquito become a major threat?
- Can armored columns be destroyed by small-unit swarm?

Miniscule spaces are best cleared by tiny units.

(Source: "Capture at Ar Ramadi," by Don Stivers, possible ©, from this url: http://www.nationalguard.mil/resources/photo_gallery/heritage/hires/arramadi.jpg.)

Modern Wars Are Fought at Closer Quarters

Winning wars against the world's most powerful armies still takes tactically proficient small units.

> "Big Enemies Require Small-Unit Cohesion."
> — title for review of last Posterity Press book
> by Lt.Col. Stoy, "Army Magazine," January 2023

To be advanced, a large-unit maneuver must consist of a composite assortment of tiny-element techniques. Every participant can't be doing precisely the same thing despite unique local challenges. Firepower-heavy Western forces gravitate toward open terrain where rote participation in every formation is possible. For such a formation to work elsewhere, each tiny element would have to slightly vary its method. Meanwhile, their ordnance deficient Eastern opponents take refuge in broken terrain that won't permit big enemy groups to enter quickly. That's why modern wars are mostly fought in urban terrain. Its compartmentalization requires a myriad of small-unit penetrations.

For the Westernized forces, any technique variation between units made easy ground slower to cross. And for the Easternized

Figure 4.1: On-Line GIs During Typical "Search and Destroy" Sweep
(Source: "American Soldier, 1966," McBarron, url: https://upload.wikimedia.org/wikipedia/commons/8/8e/The_American_Soldier%2C_1966_-_H._Charles_McBarron.jpg.)

forces, those same variations made difficult ground easier to defend. Until the West allows more leeway in small-unit technique, it will continue to lose modern wars.

There Has Been Too Little Attention to Detail in America

The separate elements of an offensive sweep must be given enough time to deal with local challenges. Otherwise, one of two things will happen. A noisy push will appear to succeed after some participants fail to notice the foe beneath their feet, and others must bum-rush an enemy gun to keep up. As in Vietnam, neither will do much for the war effort. (See Figure 4.1.)

No U.S. squad, fireteam, buddy team, or rifleman in such an operation can properly defend itself. It can neither avoid enemy fire, nor properly do its job. The squad won't search any holes in the ground. The fireteam can't outflank any machinegun. Members of a buddy team aren't taking turns covering for one other. And, the rifleman is unable to hit the deck long enough to avoid well-directed fire.

A greater tragedy is also possible . To bring to bear all available firepower, the Western outfit has risked an on-line formation that is vulnerable to long-range flanking fire. Major armies of the Eastern World have countered this threat by switching from large unit-maneuvers to the small-unit variety in series. They have replaced the linear defense with mutually supporting ambushes, and the on-line attack with a row of separate short-range infiltrations.

Western Culture Is Relatively Young

America's way of doing things comes from Europe. That's where people lived in caves around 3000 B.C., while South Asians enjoyed metropolises with wells, sewers, and public baths.[1] From that time onward, Eastern cultures continued to pay more attention to separate inhabitants and their collective opinions. But, this focus was not always humanistic. It also helped barbaric dictators to better control their subjects.

Before long, the privileged elite of Europe discovered that poor peasants crave an advocate. Kings and queens were happy to assume that role. Thus was born the "top-down" society. Such societies expected their leaders to make almost every decision without any help from below. Americans were no exception. Whenever at war with an Eastern nation, their battlefield prospects dimmed. The problem was a proficiency mismatch that U.S. civilians hadn't been allowed to correct.

Who would logically enjoy more battlefield insight—a charismatic figurehead or his subordinates collectively? But, with the coming of Mao, this ancient Asian way of solving problems from the bottom-up was ridiculed in the West. Though it had preceded Maoism by thousands of years, it was still thought to be evil. The Communist Manifesto would not be written until the mid-19th Century by two Germans—Karl Marx and Friedrich Engels.

To counteract the Western dominance in armaments, Eastern armies began to fight at very close range. They regularly defended ground so compartmentalized that tanks, artillery, aircraft, and large-unit formations couldn't seize it. That ominous task would fall to semi-autonomous Allied fireteams. Then, to defeat the latest American technologies, those mostly Asian forces moved below ground.

That's when Eastern armies began to win all wars. Their Western counterparts only managed lengthy conflicts with excess casualties (as justified through inflated artillery and airstrike estimates). By trying to maximize firepower, the West had produced to little maneuver. And as Sun Tzu had warned, "Strategy without tactics is the slowest route to victory."[2] Without careful execution, any plan can take too long.

Is American Culture Somehow Handicapped?

Western culture is not dumb *per se*. It simply lacks the grassroots knowledge of an Eastern counterpart. Nor does it make any attempt to make up the difference. For instance, U.S. leaders rarely take into account the personal beliefs, history, or subject matter expertise of separate voters. They would rather follow the advice of more powerful constituents.

The West Has Always Needed Tiny Element Activity

To beat the Japanese on Peleliu, Iwo Jima, and Okinawa, U.S.

infantry outfits were eventually forced to depend upon a few squads and their subordinate elements. While large-unit sweeps remained the official strategy, they couldn't capture many strongpoint interiors.

As at Peleliu's airfield clearing, such sweeps were often conducted within full view of elevated enemy spotters. The resulting firestorm of artillery and long-range machinegun fire greatly hurt those formations.[3] But, within the individual-award citations lies proper credit for who allowed those formations to proceed. The average U.S. headquarters had already shifted most authority to frontline units.[4] Then, pairs of volunteers took over.

On Iwo Jima, low ranking volunteers covertly skirted enemy machineguns to drop grenades into every hole on the way home.[5] On Peleliu, it took a pair of crawling infantrymen to silence the sideways-facing 30mm cannon that had wreaked havoc on most landing craft.[6] And at Okinawa, tiny teams of grunts had infiltrated forward-facing defenses to disable the automatic weapons.

> At first, the Americans did not know the pillbox caves even existed and had no doctrine for dealing with them. . . . The Americans at the outset of the offensive advanced *en masse* and were mowed down in crowds by cave-hidden machine guns when they reached the reverse slopes (Inagaki, *Okinawa,* 139-140). American officers on the scene quickly developed a method for avoiding these unhappy effects, however. They would bombard the Japanese positions they faced, forcing the IJA off the surface. They would then infiltrate men in small numbers through narrow gaps in the Japanese fixed-cave fire line ("dead spaces").[7]
>
> – *Leavenworth Papers No. 18,* by Huber

Such infiltrations had probably been responsible for the most feared U.S. maneuver at Dasheki and Wana on Okinawa. What was called "horse-mounting" by the Japanese consisted of one or two GIs somehow getting above each gun cave to toss in explosives and shoot whoever came out.[8]

Modern Low-Intensity Warfare Takes Skilled Squads

4th Generation Warfare (4GW) takes into account three non-martial arenas of conflict (political, economic, and psychological/religious). But, it is also fought in a martial arena. Its terrain is now mostly urban. This takes small units to handle all the compartmentalization.

For the Current Generation of U.S. War Heroes

There will be no attempt to skip any tactical contributions to the body of knowledge by U.S. veterans of Iraq and Afghanistan. Innovative urban and mountain techniques undoubtedly occurred at Fallujah and the Parrot's Beak. But, they weren't widely practiced, nor well explained in the after-action reports. And participating GIs would now find little mention of them in any chronicle or manual. Overly tall Western bureaucracies eat lessons learned, while mostly perpetuating the *status quo*.

So, this book must necessarily concentrate on advanced tactical techniques from previous—and better publicized—conflicts. Every whole is still the sum of its parts. And many of those parts have been around a long time. So, the relative sophistication of each technique will have little to do with the era practiced. After analyzing Japanese tactics on a large island with ample room for maneuver in 1942, Guadalcanal's commander identified the warfare style as "unconventional" and told the Marines back home how to prepare for it.

> My message... to the troops training for this type of warfare is to go back to the tactics of the French and Indian days.[9]
>
> — Maj.Gen. Alexander. A. Vandegrift

Part Two

Past Wars

"[In America's traditional style of war],
victory does not depend so much on military competence
as on sheer superiority of numbers in men."
– "Warfighting," *FMFM 1*

(Source: "Warfighting." FMFM 1.Washington, D.C.: HQMC, 1989.)

5 The French and Indian Wars

- What made the French and Indian Wars different?
- Were its tactics European or native American?

Gen. Braddock's expedition had been badly mauled in 1755.

(Source: Wikimedia Commons, with image designator "FrenchAndIndianWar.jpg.")

A Different Perspective Will Be Necessary

The average reader must now force themselves to think "small" and "historically." Otherwise, the ongoing shortfall within the U.S. military—why it no longer wins its wars decisively—will never become apparent. The generalities of high-level analyses must be set aside. In their place must be a closer look at the "basics" with which U.S. riflemen have functioned throughout the 19th or 20th Century—"shoot," "move," and "communicate." But, each should have been a category with multiple options. Victory in the colonial days of America was achieved through "micro-basics."

Alliances of the Northeastern Tribes

During the French and Indian Wars, the Hurons and Mohicans fought for the French, while the Iroquois League sided with the British colonies in America. (See Figures 5.1 and 5.2.)

> Iroquois League, . . . [was a] confederation of five (later six) Indian tribes across upper New York state that during the 17th and 18th centuries played a strategic role in the struggle between the French and British for mastery of North America. The five original Iroquois nations were the Mohawk . . . , Oneida . . . , Onondaga . . . , Cayuga . . . , and Seneca. . . . After the Tuscarora joined in 1722, the confederacy became known to the English as the Six Nations and was recognized as such at Albany. . . .
>
> During the formative period of the confederacy about 1600, the Five Nations remained concentrated in what is now central and upper New York state, barely holding their own with the neighboring Huron and Mohican . . . , who were supplied with guns through their trade with the Dutch. By 1628, however, the Mohawk had emerged from their secluded woodlands to defeat the Mohican and lay the Hudson River valley tribes and New England tribes under tribute for goods and wampum. The Mohawk traded beaver pelts to the English and Dutch in exchange for firearms, and the resulting depletion of local beaver . . . drove the confederacy members to wage war against far-flung tribal enemies in order to procure more . . . beaver. In the years from 1648 to 1656, the confederacy turned west and dispersed the Huron, Tionontati, Neutral, and Erie tribes. The Andaste succumbed to the confederacy in 1675, and then various eastern Siouan allies of the Andaste were attacked. By the 1750's most of the tribes of the Piedmont had been subdued, incorporated, or destroyed by the league.[1]
>
> — *Encyclopedia Britannica,*
> s.v. "Iroquois Confederacy"

Opposing Tribes Operated the Same Way

All native Americans vied with wild animals for food. So, their

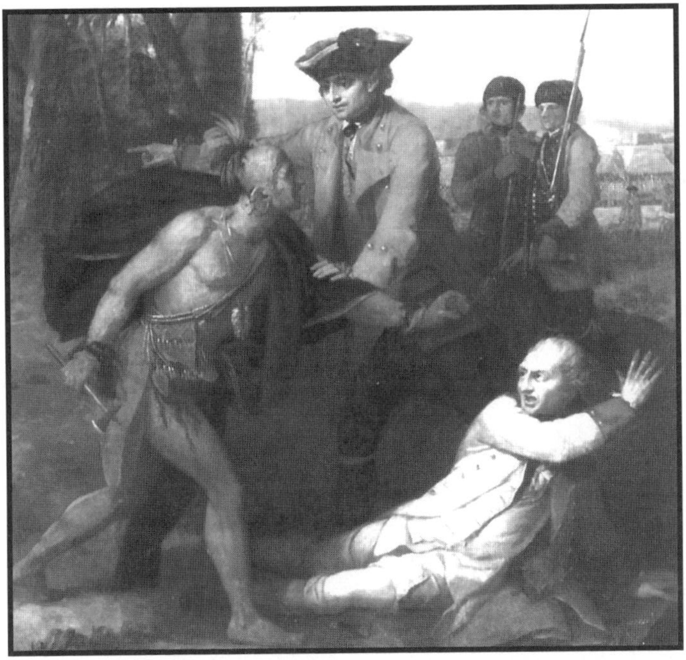

Figure 5.1: French Commander's Scalp Saved at Lake George
(Source: Wikimedia Commons, public domain, image designator "Johnson_saving_Dieskau.jpg,"

young men became as good at fighting as at hunting. Both took careful reconnaissance, patient stealth, knowledge of the ground, and surprise in the attack.[2] The Ottowas of upper Michigan, and later Ohio, would also help the French against the British colonies. Explorer Antoine de la Mothe Cadillac made very interesting record of their battlefield habits. "They go warily, . . . keep silence, observe everything, and never shoot firearms."[3] They are expert at reading sign and not leaving any behind. Once an enemy camp is spotted, they quietly study it by night and debate how best to take it with minimal casualties. They lay down on the ground, with outposts to guard against counterattack, and wait for the dawn. "Their custom is to drag themselves along the ground like cats and approach within pistol shot." They rush in and then retreat at a fast pace [and piecemeal] to their own village.[4] Like modern-day East Asian soldiers, they try to blend in and cooperate with nature.

Figure 5.2: Braddock's Men Are Ambushed
(Source: Wikimedia Commons, public domain, image designator "Destruction_de_l'armée_de_Braddock_en_1755.jpg.")

The Ottawa . . . originally lived . . . in eastern Ontario and western Quebec. . . . Their historic homelands also included . . . what is now Michigan's Upper Peninsula. The Ottawa moved into northern Ohio around 1740. They spoke an Algonquian language; and are thus related to the Delaware . . . , the Miami, and the Shawnee. Historically, the Ottawa were enemies with the Iroquois nation, and with the Wyandot. . . .

. . . Some Ottawa were allies of the French until British traders moved into the Ohio Country in the early 1700's. Many Ottawa moved into northern Ohio so that they could participate in the fur trade with the British. . . . [B]ut the British were not content just to trade. Unlike the French, the British wanted to build forts and towns.

Pontiac was a famous Ottawa leader. In 1763, he led a number of American Indians across several tribes in a resistance effort, in attempt to drive the British from American Indian-held lands. The so-called "Pontiac's Rebellion" destroyed nine out of eleven British forts in the Great Lakes region. The American Indian coalition united under Pontiac, however, could not defeat the strong British forts at Detroit (Fort Detroit) and Pittsburgh (Fort Pitt) [formerly Duquesne]. Pontiac's Rebellion came to an end after Colonel Henry Bouquet led a large army from Fort Pitt into Ohio to force the American Indians to make peace.[5]

— "Ottawa Indians," Ohio History Central

Rogers' Rangers Liked the Indian Way of Fighting

Early American settlements established their own paramilitary groups to patrol surrounding areas and guard against Indian attack. Born at Massachusetts in 1731, Robert Rogers was a frontier soldier who raised and commanded a militia that came to be known as "Rogers' Rangers." It would scout and gather intelligence for Britain's Provincial Regiments during the French and Indian Wars of 1754–63. Rogers' 600 frontiersmen had easily adapted to the Indian style of fighting, but the British Redcoats were too highly organized to grasp its potential. (See Figure 5.3.)

[Gen. Braddock, commander of all British troops in America]

undertook. . . to attack the French-held Fort Duquesne (now Pittsburgh) . . . in an extremely arduous wilderness expedition [of 1755]. . . . George Washington, then lieutenant colonel of the Virginia militia, was . . . under his command. Braddock's force safely crossed the Monongahela River and reached a point only 8 miles (13 km) from Fort Duquesne. The forward column of 1,459 officers and men, being short of Indian scouts, was ambushed in a ravine by 254 French and 600 Indians on July 9.[6]

—*Encyclopedia Britannica*, s.v. "Edward Braddock"

Figure 5.3: Rogers' Rangers Were Trained Like Indians
(Source: A painting of Rogers Rangers, "To Range the Woods, New York, 1760," by Army Artist Team XXII, iU.S. Army Ctr. of Mil. Hist., public domain)

> British officers trained in European tactics were ill prepared for the guerrilla style of fighting practiced by the French and their native American allies.[7]
>
> — *Robert Rogers' Rules for the Ranging Service* by Wulff

At first, Rogers' Rangers raided the French supply lines in upstate New York. They then conducted long-range reconnaissance patrols (largely at night). Just as the Selous Scouts of Rhodesia in the 1970's,[8] those Rangers had gained enough self-discipline—through exposure to the wild—to do well in mortal combat. After being the lead element of several British expeditions, they were asked to train British troops. At no time did they split up to function as scouts for individual Provincial units. That role normally fell to the Mohawks.[9]

Rogers' Rangers maintained a constant state of readiness throughout their existence. To help, they had *Robert Rogers' Rules of Ranging*. Those rules wouldn't protect them from ambush, but did contain enough tactical detail to avoid a massacre. One of the Rangers' more difficult outings would come against the Ottawas near Fort Detroit. It was clearly to demonstrate that a "maneuver warrior" must be as good at moving backwards as forwards.

> During the French and Indian War, Rogers took part in Gen. James Wolfe's expedition against Quebec and in the Montreal campaign of 1760. Afterward he was sent by Gen. Jeffrey Amherst to take possession of the northwestern posts, including Detroit. Rogers was again in the West in 1763, during Pontiac's War, and he participated in the Battle of Bloody Bridge.[10]
>
> — *Encyclopedia Britannica,* s.v. "Rogers' Rangers"

> In 1763, [Robert] Rogers recruited several volunteers for the reinforcement of [Fort] Detroit commanded by James Dalyell . . . . Upon arrival at Detroit, . . . [they joined] Dalyell . . . [in an] attack [on] an Indian village near Parent's Creek. The force of 250-300 soldiers . . . , Rogers' volunteers, and the Queen's Royal American Rangers . . . [were] ambushed, as the advanced guard . . . crossed the bridge at Parent's Creek [soon to be called the Battle of Bloody Creek (or Run)]. Rog-

> ers' men were responsible for effectively covering the retreat of the force back to Fort Detroit.[11]
>
> — *Wikipedia Encyclopedia,* s.v. "Rogers' Rangers"

> At the Battle of Bloody Run, a fierce fight broke out at the bridge, with the [enemy] Indians fighting under cover and darkness of the night (the march had begun at 2 A.M.) The [British] Regulars, unable to bring the attackers to close combat, were forced to retreat. . . . This fighting retreat was probably done using the street firing tactic. . . . Major Rogers had taken command of a [nearby] house . . . and by diverting the attackers attention from the main body, covered their withdrawal.[12]
>
> — *Robert Rogers' Rules for the Ranging Service,* by Wulff

Rogers' Tactical Guidelines

As longtime woodsmen, most of Rogers' men had learned how to fight like native Americans by 1757.[13] Now able to hold their own against woods-wise scouts, the Rangers were soon able to operate as a mobile whole.

But, like with hunting, fighting in the forest ultimately came down to how well each individual could do on his own. He had to be able to sneak up on an opposition encampment, win a one-on-one encounter with an enemy soldier, and then individually elude any number of pursuers.

Attention to Detail

Throughout the 20th Century, U.S. Marines also had to "pay attention to detail." During the frozen exit from Korea's Chosin Reservoir in late 1950, too few dry socks meant a non-combat-ready unit. So were the small-unit maneuvers quite demanding. For example, ambushes had to have: (1) 360° security; (2) no one forward of the firing line; (3) no pre-washing with perfumed soap; (4) all claymore mine wires retrieved; (5) etc. No officer in charge, nor the appropriate manual, knew all the details of a successful mission against the Red Chinese in 1950. Many had been learned the hard way against the Japanese and then passed along by word of mouth.

In the 1750's, if the path to an ambush site hadn't been wiped clean of all human sign, the occupying American squad could be jumped by 150 screaming Hurons. The false security of standoff warfare had yet to take hold. The twig snap of an ambusher shifting his weight could lead to disaster. So, any young GI wishing to survive the next big war should read on. Those who will never find themselves alone overseas can skip ahead to the next chapter.

Rule #1

Rogers' Rangers were never caught off-guard. They could leave on a multi-day operation at a moment's notice. All base camp and movement security billets were pre-assigned. Rogers' men carried 60 rounds so as to not require resupply. Their fire discipline was then a product of necessity. Unlike other troops of the period, they independently reloaded and fired their rifles without any leader's direction. The Rangers' self-sufficiency, courage, stealth, and camouflage were how best to survive in a heavily vegetated environment.[14]

Rule #2

By traveling single file, Rogers' patrols created a thin frontal signature that more easily dodged the security elements of an enemy base camp. The lead Ranger picked a route that would produce the least motion and sign. He recommended any hasty ambush by the main party, while flankers kept that party from being surprised.[15] Rogers often conducted a "scout" with five men or less. Sidestepping enemy concentrations would become the central theme of 3GW.

Rule #3

While crossing soft ground, Rogers' Rangers spread out abreast to lessen the amount of sign they left. Each man took responsibility for his own discernible trail.[16] To get closer to the objective and keep their bivouac secret, Rangers kept moving well into the night. They then picked a rendezvous site from which intruders could be seen from afar—like a tiny knoll or mid-clearing thicket.[17]

Rule #4

A small advance party can more easily pass through the roving outposts of an enemy camp than the main body. This was to become the principal axiom of short-range infiltration—two-man teams sallying forth from a well-defended staging area.[18]

Rule #5

Rogers valued the taking of prisoners for the intelligence they could provide. Those prisoners were kept apart to avoid shared falsehoods. Nor are tortured prisoners ever dependable. They will say anything to stop the pain. All that matters is the enemy situation. With the onset of "total war," killing became more important than discovering that situation.[19]

Rule #6

Three columns abreast advance as quickly as one. If attacked from a flank, the unit is properly aligned when all members face that direction. Then, that side's security becomes the lead element. It can be quickly expanded with personnel from the first row. Much like the VC's double-line defense in Vietnam,[20] the first row can also fall back upon the second.[21]

Rule #7

Rogers had discovered an important axiom of chance enemy contact that is still not fully appreciated. Attack immediately, no matter how big the opposition. Chesty Puller would do this on Guadalcanal and Peleliu, but be later criticized for risking too many lives.[22] Rogers had further realized that one best avoids enemy fire while prone. Then, "fire and movement" through some cover becomes much safer. But, Rogers' fire and movement was by many individuals between trees, instead of the bounding overwatch of contemporary small units. Today, even within less cover available, buddy team fire and movement is the most efficient.[23] That's when each team has its own lane, and the moving man is covered by a stationary buddy. A Ranger only fired when a member of his own fireteam had stopped to reload his weapon.

Rule #8

First engaged in a chance encounter may be the advance party of a larger unit. It should be carefully pursued and kept from rising ground. Only a friendly advance party can discover a counterattack in time.

Withdrawal in good order is part of Chinese mobile warfare; and so are well-laid traps. While European tactics stressed on-line frontal assaults, "the native Americans were well known for . . . sweeping around the flanks of an enemy and eventually surrounding [him]."[24] That flanking force then stayed behind trees and shot from a distance at their fully exposed quarry.[25] A modern-day Asian does the same thing while retreating.

Rule #9

A fighting retreat must take advantage of the microterrain. Fallen logs provide cover while swampy ground forestalls trouble from a flank. But, backwards buddy team fire and movement produces more return fire than a rearward bounding overwatch. In 18th Century "'street firing," back-to-back fireteams took turns shooting. The front team would fire and then move behind the other to reload.[26] There is no modern equivalent. But, backwards-moving fire and movement by a row of buddy teams produces the most continuous fire. Instead of separate volleys, the enemy endures a staccato stream of bullets.

Rule #10

Rogers' Rangers sometimes scattered to avoid capture. They did so individually to leave less trail. That's the same as the short-range exfiltration of an enemy encirclement. Survivors of the "Battle of Snowshoes" had fully dispersed to save their lives. Just as a rendezvous spot is established during a short-range infiltration, so must one for an exfiltration.

A square formation will defend level ground because it allows for a concentration of fire to any one side, or for one side to fall back upon its opposite.[27] In the modern-age, four machineguns can form a quadrilateral cocoon of interlocking FPLs (final protective lines).

Rule #11

At the heart of Rogers' Rules is readiness at a moment's notice. There could be no major change of formation and very few orders. A short row of columns can meet an attack from any direction by having all exterior personnel face outboard.[28] And, a unit feigning retreat must only reverse direction to counterattack.

In effect, Rogers had come up with "tactical techniques" for a company-sized unit, just as the Chinese had directed by whistle at the Chosin Reservoir in 1950.[29] At present, Rogers' insights are most useful to smaller units. A single-file patrol can easily handle a flank threat by every member facing to the right or left. They're now on line and capable of maximum fire. The instantaneous counterattack by a fleeing unit would become the favorite trap option of the "retreat combat" variant of 3GW.

Rule #12

Outside Fort Ticonderoga in March of 1759, Rogers' Rangers made a stand just beyond a ridgeline to shock their pursuers.[30] So doing may have been one of the first recorded reverse-slope defenses. When Rogers counterattacked from rising ground, he left a few people behind to allow for any change of plans.[31]

Rule #13

Rogers' Rangers were the first European Americans to prefer close combat. Their final charge was usually preceded by a musket volley. In later years, that way of fighting became necessary against a better-armed adversary—as with the "hugging" of U.S. units by East Asian soldiers. On Okinawa, the final charge was preceded by a flurry of hand grenades.[32] Lightening the Rangers' load had led to the use of hatchets and small cutlasses (instead of bayonets) in the forest.[33] In heavy jungle, a machete might get both jobs done.

Rule #14

Rogers allowed no noise on his missions. That's why his troops so easily avoided detection. They could crawl right through an opposition sentry line. While encamped, six were assigned to each listening post (LP). This permitted enough sleeping to keep those on watch alert. One man would carefully sneak back to report any sightings. No LP was relieved until dawn to preserve the secrecy of its location.[34] The Marines on Guadalcanal also found buddy team LPs too dangerous to use against a woods-wise adversary.[35]

Rule #15

Rogers preferred to attack an opposition target in the predawn hours while its defenders were still groggy, in their nightclothes, and trying to adjust to the low light.[36]

Rule #16

Rogers liked to attack during the night because it hid the size of his assault force and allowed for an easier retreat.[37] He moved in slightly after most defenders had fallen asleep.[38]

Rule #17

Each morning, Rogers' Rangers searched around their bivouac for footprints from the night before. Such prints would often reveal enemy intentions. Whether or not the opposition force was too big to pursue became quickly apparent.[39]

Rule #18

Rogers closely monitored his Rangers' liquid intake and would stop some distance from any source of water to keep them from getting ambushed. Even when pausing to eat, he put out sentry posts.[40]

Rule #19

Rogers' Rangers would backtrack before any river ford to make sure they hadn't been followed. They could always cross at a less obvious place. Such actions were the forerunner of avoiding all patterns of movement in UW.[41]

Rule #20

Rogers used stream or lake shorelines only as a navigational aid. Too closely following one relinquished a possible avenue of escape.[42]

Rule #21

To ambush pursuers, Rogers' men circled back on their own tracks.[43]

Rule #22

Rogers' people could become too lax while approaching their own fort after a long scouting mission. So, they were always well fed and rested before returning home.[44]

Rule #23

Following too closely an enemy contingent near one's own base can be dangerous. It's much better to guess where that contingent is headed and get there first.[45]

Rule #24

Rogers' Rangers often traveled at night on land and water to avoid being seen.[46]

Rule #25

A way of maintaining contact between a string of boats would also work with small units in column. Each unit stays in contact with the one behind it. Normally, human-file members only watch the soldier to their front.[47]

Rule #26

Opposition soldiers reveal their presence through campfires at night. From the number of fires, the enemy's strength can be determined.[48]

Rule #27

Lingering along the opponent's assumed avenue of retreat is an excellent way to ambush what remains of his attacking contingent.[49]

Rule #28

Determining an enemy's actual strength often takes a reconnaissance party. Countersigns must be issued and rally points assigned.[50]

Half of the Planet Would Lose Interest in Footprints

Most French and Indian War participants became so good at reading human sign that they could predict enemy intentions. But, throughout the Western World, the focus was about to shift from opposition capabilities to friendly unit functioning. Only East Asian armies would maintain much tracking or counter-tracking proficiency. For example, how Ho Chi Minh Trail construction crews kept their project secret during the Vietnam War. While traveling single file, the lead man gently separated the leaves and grasses with a stick, and the last man walked backwards to restore them to their original condition.[51]

Did GIs No Longer Need Such Skills?

All the wonderful equipment that was soon to burden Western infantrymen would not make their evidence of passage any less obvious. Nor would it change what might happen to a sentry post or scouting team from behind that had just "snuck" out onto an active battlefield.

Tracking and counter-tracking are just as important now, as they ever were. And, they are only two of dozens of subcategories of the U.S. basic "to move" that the average rifleman never hears about. Both are quite complicated. Those who will send those trusting young souls into harm's way should take a few moments to appreciate their details.

Current (Albeit Unshared) Knowledge of Counter-Tracking

There are many ways to foil a tracker. They fall into four general categories: (1) increasing the time/distance gap; (2) disguising/concealing sign; (3) reducing sign; and (4) counterattacking.[52] Least efficient is the first.

Quickly covering more ground creates hard-to-miss evidence of passage that an experienced tracker can easily use to leapfrog ahead.

Footprints can be concealed or disguised in many ways. (See Figure 5.4.) An area can be thoroughly brushed out. But then, it will be devoid of all animal sign and weathering. When a leafy branch is used too long for this purpose, it turns into a twig bundle that produces unmistakable scratch marks. By following in each other's footsteps, those being pursued can disguise their number. Their column can follow (or offset along) hard ground and streambeds.

But, the latter may create underwater footprints, crushed/broken water plants, unnaturally wet rocks, and bank-side skid marks. (See Figure 5.5). Muddy water alone will alert anyone downstream to the ruse. The last man in column can restore displaced vegetation to its natural state with a stick. He can cover his and his comrades' tracks with leaves. Or he can drive large numbers of people or animals over the tracks. Initially choosing a heavily traveled trail accomplishes the same result. Then, various foot coverings will disguise the sole pattern, but also add width to the imprint. Among the most ingenious are leaves. Shoes can be removed or interchanged with wide pads and animal feet. Another trick is to walk backwards. But, as the foot's initial strike shifts to the toe, soil particles will reveal the actual direction of movement. On soft ground, the heel print will also be deeper and the pace shorter.[53]

Reducing sign can be equally multi-dimensional. The lone quarry can step from rock to rock. He can follow roots, fallen trees, rocky ground, or paved road. Or, he can place his weight on harder ground at the base of each bush. If he tiptoes, there will be no heel marks. He can suddenly veer off to one side. A partial reversal of direction works best. Even an experienced tracker will lose valuable time looking backwards for the next print. Then, there are composite techniques. The quarry might move past a tree, retrace his exact footsteps, and then jump to the blind side of that tree (a method used by some animals).[54] Or he might move 20-30 meters back along the stream he just came up and then carefully exit through a feeder stream or thick brush. There are also ways to minimize human sign. The group can break up into subgroups. Or every member can take off in a different direction to rendezvous later. Individuals can drop off at intervals, using rocks or sun-baked ground to exit without a trace. Every other man can carry his companion piggyback. This causes the tracks to get deeper. A lone quarry can crawl off on his hands and knees or roll along the ground. Yet, he will leave behind hand prints, boot toe indentations, and equipment drag marks. He can swing off the trail on a vine or low-hanging limb. This will leave telltale clues—a sudden disappearance of tracks, broken branches, and missing or scuffed bark.

Finally, the pursuing tracker can be discouraged or eliminated altogether through some form of counterattack—sniper, boobytrap, or ambush. In Haiti and Nicaragua, a Chesty Puller patrol would drop off a small welcoming party.

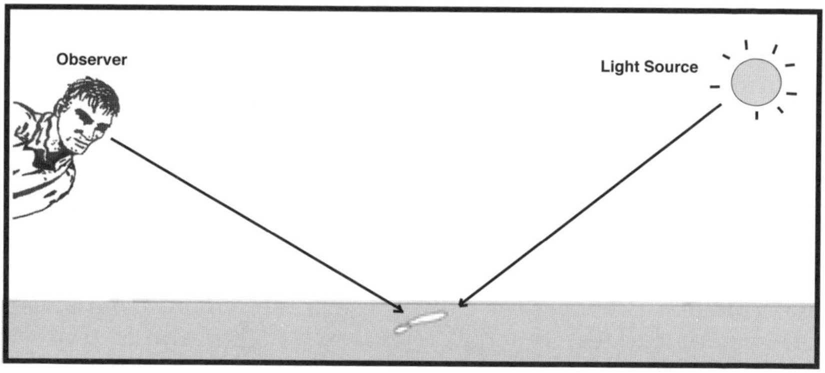

Figure 5.4: Optimal Viewing Conditions
(Source: *FM 21-76* (1957), p. 38)

Was the First Example of 3GW in America?

The native Americans of the French and Indian Wars would "flow around" and encircle a enemy force to decimate its ranks from the cover of surrounding trees.[55] So doing obeyed not only the laws of nature, but also East Asian advice from 500 B.C.

> An army may be likened to water, for just as flowing water avoids the heights and hastens to the lowlands, so an army avoids strength and strikes weakness.[56]
> — Sun Tzu

A hundred years later on the Great Plains, settler or army contingents were similarly surrounded before any penetration. Even when mounted, Western Indians followed the same pattern. Their base of fire was from every direction at once; and the assault occurred at the weakest point of the perimeter. In effect, several generations of native Americans had focused more upon their adversary than their own internal interaction. This is the principal difference between MW (3GW) and 2nd Generation or Attrition Warfare (2GW).[57]

The French of mid-18th Century America are said to have been practicing Guerrilla Warfare (GW)—a subset of 3GW.[58] But not until Napoleon invaded Spain in 1808 did the French Army learn much about the modern version of GW.

In the deep woods of upstate New York, neither side would come fully on line during a chance encounter. The advance party did to some extent, and its flankers as well. But, any all-out clash generally involved a loosely controlled mass of soldiers, moving from tree to tree, and then individually engaging an opposition fighter. The only European maneuvers were a defensive square and the outgoing fire of successive ranks in column. Everything else was of Indian origin. Rogers' Rangers ability to quickly disperse and then reassemble would not be among the German parameters of MW in 1917.[59] It would be later added by the Chinese Communists. While fully consistent with MW, the idea had long existed in Mainland China.

Figure 5.5: Running Water Won't Obscure All Sign
(Source: *FM 21-76* (1992), p. 8-27)

> The Cloud Battle Array is used when both sides are shooting at each other with bows and arrows [or indirect fire]. Such a battle array is flexible in making changes and gathers and disperses troops quickly.[60]
>
> — Sun Bin (316 B.C.)

And, the native American way of war had been like that of East Asia in other ways. On Operation Buffalo in Vietnam in 1967, while moving north along Route 561, Company B, 1st Battalion, 9th Marines walked into the pre-arranged encirclement of a "U-shaped" *Haichi-Shiki* from Mao's Mobile Warfare (3GW) portfolio.

A single sniper had opened up on the Marines. Then, comrades in a trench 100 meters to the west of Route 561 started taking pot shots at the Americans. Already on line facing that direction, a U.S. platoon assaulted what they thought was an enemy squad and got cut up by machineguns. Then, to the dismay of assault survivors, a sizeable North Vietnamese Army (NVA) force advanced toward Route 561. Instead of a human wave, it did so as a ragged line of occasionally moving "grass clumps."

Meanwhile, Route 561 had been under intermittent mortar and machinegun fire from the north. And NVA soldiers manning a belt of spider holes along the road's right side had started to pick off exposed Marines. Then, a third NVA force moved down from Hill 70 (south of the beleaguered column) to close off the bottom of the inverted U.

Of note is how foe detachments had kept from shooting each other. Most effective had been the random shots from spider holes east of the road. They were hard to hear above the western cacophony. By occasionally firing, the spider hole occupants had kept their locations secret. The NVA west of the road used other tricks. They shifted location between shots. They moved toward the road in short three-man rushes. And while stationary, they remained indistinguishable from the surrounding vegetation.

When the Asian Communists got close the road, they altered their routine. Riflemen separately crawled and randomly fired. By taking well-aimed shots, they kept from hitting their dug-in brethren across the road. And, some went to even greater lengths to avoid fratricide.

> There were NVA climbing up in the trees to fire down on

> the pinned-down Marines from 1st Platoon and . . . 2nd Platoon.[61]
>
> — *Operation Buffalo,* by Nolan

Under intermittent fire from all sides and unable to acquire good targets, the Marines became disoriented. That's when NVA fireteams crossed the road at several places to segment the column.[62] And the annihilation attempt continued piecemeal. Marine reinforcements eventually arrived and most of the men of B/1/9 were saved.

But, a few questions linger with regard to the French and Indian Wars. Why had descendents of those who crossed the Alaskan land bridge thousands of years ago fought like the Asian Communists of the modern era? Was it instinctive behavior, or frontline common sense that had never been suppressed?

More Evidence of 3GW in Mid-18th Century America

Were Rogers' Rangers and their Indian role models really practicing 3GW that long ago? Only two indicators of this startling conclusion have so far been noted. Their mutual disregard for unit dynamics was evident from the delegation of authority to front-line fighters. And, both used encirclement ("entrapment" in 3GW terms).[63] To prevent fratricide, 2GW forces are seldom allowed to double-envelop a foe.

> Rogers . . . [had] admonished that most of what he had learned was subject to change according to, ". . . [local] circumstances." . . . Then he stated: "every man's . . . judgement must be his guide . . . on every occasion." This statement was revolutionary in that he gave ordinary . . . [soldiers] the freedom to think for themselves.[64]
>
> — *War on the Run,* by Ross

But, there are other aspects of Rogers' method that make it unmistakably 3GW. Among them are speed, surprise, and very little shooting. A 2GW push is slow, methodical, and heavily supported by firepower.[65] The most telling of the Rangers' procedures was falling back in the face of a stronger enemy force. 2GW forces won't do that.

Not only would Rogers' Rangers retreat, but occasionally scatter in an "every-man-for-himself" skedaddle. At least, that's what happened at the Battle of Snowshoes. Meanwhile, European-trained soldiers were forced to hold their ground under every circumstance. Fully trusting what their leaders had told them about close combat, they found themselves out-matched by a survivalist Indian or Ranger.

Tactical withdrawal has since become a perfectly legitimate aspect of warfare. Toward the end of WWII, retreat combat (a subset of 3GW) would be heavily practiced by both Germans and Japanese.[66] In the process, many an Axis pocket would be successfully "exfiltrated" by its occupants.

A Strongpoint Thread Will Be Necessary

All subsequent U.S. wars will now be studied to determine what occurred at the lowest echelons of the average attack. That's where casualties are most easily avoided, and technique refinements apparent. In the French and Indian Wars, any serious resistance had been bypassed, encircled, and then carefully eliminated. As in the Motti ambush of the Soviet-Russian War and *Hachi Shiki* of the Korean or Vietnam Wars, this is a manpower thrifty (and thus advanced) way of fighting.[67] At the end of each chapter, how GIs captured that war's strongpoints will be assessed for safety. So doing should help the reader to see how much tactical progress had been made from the preceding conflict.

6 U.S. War of Independence

- Did Colonists ever fight like Indians during the Revolution?
- How many times?

British column on its way back to Boston from Concord.

(Source: Wikimedia Commons, with image designator "Ross_Moffett-_A_Skirmish_Between_British_and_Colonists. . . Smithsonian_American_Art_Museum.jpg.")

During the First Revolutionary War Encounter

The preceding chapter has demonstrated that "ambush" and "infiltration"—the future fundamentals of 3GW variant UW [1]—had been widely practiced in the French and Indian Wars. In fact, most settler militias had tried to emulate the Indians' way of fighting. Then, during the first battle of the Revolutionary War in 1775, indigenous peoples were present to make sure the Minutemen shot their adversaries from cover after first surrounding them. But, before long, too much organization would saddle the Colonials' effort with inertia.

Initial Reality Check

In war, that which hastens victory is productive, and other things not so much. Yet, many competing agendas come into play—like pride, tradition, vengeance, untested equipment, and unit discipline.

Shortly after American Minutemen embarrassed a British expeditionary column—the Indian way—at Lexington, leaders of the mostly English immigrants made a poor decision. They felt somehow obligated to beat the Redcoats at their own game. The assumption was that sedentary Europeans had become smarter than migratory Asians. Those erstwhile members of a top-down society had yet to realize that bottom-up societies existed, must less excelled at everyday problem solving.

> The Stockbridge Mohicans went on to fight in some of the earliest battles of the war and were reportedly among the militiamen at Lexington and Concord in April of 1775, according to Rindfleisch.
>
> Mohican warriors are said to have lined up alongside American militiamen along Battle Road and fired upon British soldiers as they marched back to Boston and then joined the militiamen in Cambridge as they besieged the British army within Boston.[2]
>
> — *History of Massachusetts Blog*

Then, when 2400 British regulars took Bunker (Breed's) Hill on their third assault after 1000 American defenders had run out of ammunition in June 1775, Colonial war planners became even more heritage oriented. The Redcoats' fatalistic courage had so impressed them that they wanted their own soldiers to withstand volley musket fire. No longer legitimate were the clouds of lightly armed fighters silently sweeping through the woods to surround and gradually annihilate any opposition. In their place must be orderly rows of utterly fearless blue-clad troops. The Western leaders' predisposition toward bringing order to chaos had won out over frontline common sense. But, then, it was not they who would do the dying.

The forts of the French and Indian Wars had largely depended upon those battles beyond their walls. But, after Bunker Hill, their rampart design and cannonball supply seemed more important. It had been too early to recapture Boston without enough ammunition, so the strategic value of Bunker Hill had been questionable from the outset. And, whether the Indian or European way of fighting had been more "evolutionary" is now painfully clear to students of infantry tactics. Failing to sneak up on a prepared enemy position entails considerable sacrifice. The measure of excellence for any small-unit maneuver is how many friendly casualties it precludes.

For the Military Professionals

How smart is it to dress in red, stand shoulder to shoulder, and then march up a contested hill? Against volley musket fire, it may eventually succeed, but only after heavy losses. The Bunker Hill episode would cost the British 1000 killed and wounded, versus 400 for the Americans. The other 600 Americans must have withdrawn after hand-to-hand fighting, because only 30 were captured.[3] But

Figure 6.1: British Had Taken Terrible Losses at Bunker Hill
(Source: Wikimedia Commons, public domain, image designator "Battle_of_bunker_hill_by_percy_moran.jpg,"

such rearward movement had been justified. How helpful is mass surrender? Rogers' Rangers would have dispersed prior to the third British assault, and never had their bravery questioned. But falling back for any reason became so distasteful to Americans, that it would prevent them from fully practicing MW well into the 21st Century.

There were exceptions, of course. On Francis Marion's advice in the movie "The Patriot," Colonial militiamen pretend to break and run at the Battle of King's Mountain, only to lead the British into a counterattack-allowing trap.[4] All war is based on deception according to Sun Tzu,[5] and methodically (though bravely) moving forward is too predictable to be wise. Mental toughness must be combined with common sense or a bunch of people start wandering around within full view of the enemy.

Extent of Indian Participation in the Revolution

Of the six Iroquois Nation tribes, only the Oneida and Tuscarora ended up siding with the Americans. The Mohawk, Seneca, Cayuga, and Onondaga all helped the British—probably as scouts during the ill-fated American forays into Canada. Meanwhile, the North Carolina Tuscarora may have scouted for the Colonials' more successful Southern initiatives.[6] But, cautiously fighting like an Indian was to be largely replaced by the more dangerous European approach. Not until Yorktown would the gradually tightening Indian cordon reappear.

In New England, the Penobscot and Pasmaquoddy tribes sent 600 men to join the Continental Army. Some of the Penobscot warriors then served as scouts during the Canadian expedition. And a whole regiment of Stockbridge Mohicans took part in several battles. Besides Lexington, there were some Indians at Bunker Hill. And Oneida tribesmen were at the crucial engagements of Orisanki and Saratoga.[7]

Within the Great Lakes region, the Potawatomi and Delaware sided with Americans, while the Huron (Wyandot), Shawnee, Miami, and Ottowa aided the British.[8]

> During the American Revolution, the Ottawa fought for the British against the Americans. When the British surrendered to the Americans, the British turned their backs

> on their American Indian allies. The Ottawas continued to fight the Americans.[9]
>
> — Ohio History Central, s.v. "Ottawa Indians"

In the Southern region, the Catawba and Tuscarora supported the Americans, while the Chickasaw, Choctaw, Creek, and Cherokee aided the Redcoats.[10]

Why Did Early Colonials Fight as Individuals?

Most warlike of the Iroquois League tribes may have been the Seneca. They regularly raided far south along the Allegheny Mountain chain. But, the Seneca believed one member of the tribe should never impose his will upon another.[11] Like the Ottawas during the French and Indian Wars, the Seneca may have fought individually and then withdrawn separately.[12] So doing is an advanced concept, the true value of which would not be realized by U.S. infantry commanders until late in WWII.

> The Indian method of warfare . . . [inevitably] adopted by the white man also, was the most significant influence in developing . . . individualism and self-reliance in the military sphere. When the white man came, the Indian . . . soon learned the value of the white man's muskets and was not long in obtaining them. . . . With bow or musket, his method of fighting was the same. Indian tribes had no organized system of war; warriors simply formed voluntary bands under war chiefs and took off on the warpath. In battle, each Indian fought a separate opponent without regard for his fellows. Indians avoided pitched battle whenever possible, instead seeking victory by surprise and carefully utilizing cover and concealment. Only when they had the advantage did they close in for hand-to-hand combat.[13]
>
> — "The American Revolution," *CMH_Pub_70-6-1*

But, what had been an Indian strength emulated by English immigrants was soon deemed a weakness. The settlers' pride-filled heritage had dulled their sense of reason. How could the British Empire have been created without the world's best infantrymen?

So, initiative—one of the most important aspects of short-range combat—would be discouraged in most American recruits from then on.

> The American soldier was inclined to be highly individualistic and to resent discipline and the inevitable restrictions of military life; he sought to know why he should do things before he would put his heart into doing them.[14]
>
> — "The American Revolution," *CMH_Pub_70-6-1*

Summary of the American Revolution

The U.S. War of Independence would go on for eight long years. Some battles helped the war effort more than others. But, as a whole, they forced the British into coastal areas where a decisive encirclement was possible. Still at war worldwide, the British would eventually have to abandon their American intentions.

> Fighting began outside of Boston in the spring of 1775 during a British raid to seize munitions at Lexington and Concord. British regulars arrived on the Lexington Green early on the morning of April 19 and discovered the town's militia awaiting their arrival. The "Minutemen" intended only a show of force, and were dispersing, when a shot rang out. The American War of Independence had officially begun.
>
> The militia harassed the British all the way from Concord to Boston, and then surrounded the city. In an attempt to drive the Colonials away from the city, British forces attacked the Americans at Breed's Hill on June 17th, resulting in heavy casualties for the Redcoats in the war's first major battle. George Washington arrived that July to assume command of the American forces, organized as the Continental Army. Washington then forced 11,000 British soldiers to evacuate Boston the following March, when Henry Knox successfully led 12 artillery pieces from Fort Ticonderoga to Dorchester Heights overlooking the city below.
>
> By the early spring of 1776, the war had expanded to other regions. At Moore's Creek in North Carolina and

Sullivan's Island at Charleston, American forces stopped British invasions. After initial successes, particularly the capture of Fort Ticonderoga in upstate New York, an American invasion of Canada stalled and ended in failure at the end of the year. As 1775 rolled into 1776, the British rapidly built up forces in New York and Canada to strike back.

After a series of five consecutive defeats for Washington's army at Long Island, Harlem Heights, White Plains, Fort Lee, and Fort Washington, the British captured New York City in the summer of 1776. Following the capture of the city, the British drove Washington's army across New Jersey, winning several additional battles along their advance. That winter, however, Washington revived the American cause by winning spirited victories at Trenton and Princeton, New Jersey. [See Figure 6.2.]

In 1777, the British launched two major offensives. In September, General William Howe captured Philadelphia, winning battles at Brandywine and Germantown. Despite the losses, the inexperienced soldiers of the Continental Army performed well and gained a measure of confidence, believing that they could very well stand up to the British. Then, in October, British General John Burgoyne invaded upstate New York via Canada, winning several initial victories. Later, however, his army became bogged down thanks in part to efforts of American militia units at Oriskany, Fort Stanwix, and Bennington. Then, after a stunning defeat in an open battle, Burgoyne surrendered his entire field army at Saratoga, New York.

The American victory at Saratoga was a turning point of the war, for it convinced the French monarchy that the Americans could actually defeat the British in battle. As a result, a formal military alliance was signed between the French and American governments in 1778, which entailed increased financial and military support. The alliance had even more positive implications for the Continental Army, because it forced the Parliament to funnel manpower and resources to fight the French across the globe, rather than sending them to North America.

That same winter, a few months prior to the formal signing of the alliance, Washington's army retired to Valley Forge, not far from the British garrison in Philadelphia. While arriving rather disheveled, disheartened, and largely undisciplined, the army underwent a rigorous training program under the direction of Baron von Steuben. He instilled in the soldiers a sense of pride, resilience, and discipline, which transformed the army into a force that was capable of standing toe-to-toe with the British.

In 1778, the British consolidated their forces in New York and Canada and prepared to launch an invasion of the South. In the meantime, in the west, American forces under George Rogers Clark captured several British posts, culminating with a victory at Vincennes, Indiana, and the surrender of a much larger British force.

To the North, the British abandoned Philadelphia for New York with Washington hot on their heels. His army caught up to the Redcoats at Monmouth, New Jersey, where an intense battle ensued. After arriving late to the battle and rallying his wavering troops, Washington made several

Figure 6.2: The U.S. Assault Line at Trenton
(Source: Wikimedia Commons, public domain, image designator "Henry-revolutionary-war.jpg,"

defenses and counterattacks against the surging British force. Though inconclusive with no clear victor, the battle demonstrated the growing effectiveness of the Continental Army. Upon finally reaching New York, British forces never again ventured far from their secure base there.

In 1779, with . . . a stalemate developing in the North, the British began to focus their efforts on conquering the South, in hopes of quelling the rebellion once and for all. That autumn, British forces captured Savannah and Charleston and smashed General Gates' army in Camden, South Carolina. . . . However, the Continental Army won battles at King's Mountain and Cowpens, stemming the tide of British advance. Undeterred, the British army under General Charles Lord Cornwallis then moved across North Carolina before fighting its way into Virginia.

While General Cornwallis fought his way into Virginia, a brutal civil war erupted among the civilian population of the Carolinas. General Nathanael Greene recaptured most of South Carolina, fighting battles at Ninety Six, Hobkirk's Hill, and Eutaw Springs. While Greene lost most of the battles in which he fought, he skillfully used his mixed force of militia and Continental regulars to maneuver the British out of the Carolinas' interior, forcing them toward the coastal cities and towns.

By the summer of 1781, Virginia was ablaze with battles along the colony's coast and across its center. As General Marquis de Lafayette doggedly forced Cornwallis toward the coastal defenses around Yorktown, Virginia, he persuaded Washington to move the Continental Army from Connecticut to Virginia. Washington, along with a French fleet and army commanded by General Rochambeau, arrived in Virginia on September 19th, 1781, effectively sealing shut any escape route for Cornwallis. Following a siege and a series of attacks on the British position, Cornwallis surrendered his army to Washington.[15]

— *American Battlefield Trust*

America's "Revolution" Mostly Wasn't

In the 18th Century, revolts normally occurred through small-unit rebel activity. But, only in the American South would the U.S. War for Independence take on the characteristics of GW. Elsewhere, there were piecemeal battles of attrition. And, the structure of positional warfare seemed to appeal more to the Continental Army than the free-wheeling successes of mobile warfare.

> While the American Revolutionary War is often thought of as a guerrilla war, guerrilla tactics were uncommon, and almost all of the battles involved conventional set-piece battles. Some of the confusion may be due to the fact that generals George Washington and Nathaniel Greene successfully used a strategy of harassment and progressively grinding down British forces . . . instead of seeking a decisive battle. Nevertheless the theater tactics used by most of the American forces were those of conventional warfare. One of the exceptions was in the south, where the brunt of the war was upon militia forces who fought the enemy British troops and their Loyalist supporters, but used concealment, surprise, and other guerrilla tactics to much advantage. General Francis Marion of South Carolina, who often attacked the British at unexpected places and then faded into the swamps by the time the British were organized enough to return fire, was named by them "The Swamp Fox."[16]
>
> — *New World Encyclopedia,*
> s.v. "Guerrilla Warfare"

Early Meaning of GW

The French and Indian Wars are said to have been fought through GW. But, the term's meaning was different then.

> The early years of the French and Indian War were marked with . . . [t]he resounding defeat of . . . Braddock's Army at the Monongahela River. . . . British officers trained in European . . . tactics were ill prepared for the guerrilla style of fighting practiced by the French and their native American allies.[17]
>
> — *Robert Rogers' Rules for the Ranging Service,*
> by Wulff

While wary of GW, the British still considered it less legitimate than more conventional ways of fighting. During the American Revolution, many Colonial leaders had been veterans of the French and Indian Wars. George Washington, Israel Putnam, Philip Schuyler, and John Star had served with the Provincial forces, while Charles Lee, Horatio Gates, and Richard Montgomery fought for the British Army.[18] Even among those veterans, GW was widely believed to depend for its success upon dense woods. In more open terrain, its reliance on the initiative and common sense of the individual would fail to produce the cohesion and discipline of entire units.

> In the Indian wars and the colonial wars with France, Americans gained considerable military experience, albeit much of it in guerrilla warfare that did not require the same degree of organized effort and professional competence as the European style of warfare.[19]
>
> — "The American Revolution," *CMH_Pub_70-6-1*

Having first appeared near the end of the French and Indian Wars, the Kentucky rifle had been good for GW. But, its use from a distance wasn't considered as "manly" as fighting out in the open at close range.

> For the guerrilla tactics of the frontier, where men did not fight in line but from behind trees, bushes, and rocks, it [the Kentucky rifle] was clearly a superior weapon.[20]
>
> — "The American Revolution," *CMH_Pub_70-6-1*

Each Kentucky rifle had to be handmade, so spiral grooves in a long barreled small arm wouldn't become standard issue in America for 100 more years.[21]

The Swamp Fox's Version of GW

Francis Marion had witnessed good small-unit tactics in the Anglo-Cherokee War of 1758-1761.[22] He then applied a strategy of gradual depletion in the Colonial War. Instead of decisive victory against a major force, he raided convoys and outposts.[23] From such raids, he seemed more powerful than he actually was.

> Marion's tactics were those of all successful guerrillas. Operating with the greatest speed from inaccessible bases, which he changed frequently, he struck his blows in rapid succession at isolated garrisons, convoys, and [supply] trains.[24]
> - B.Gen. Samuel B. Griffith USMC (Ret.)
> former CO of 1st Raider Battalion
> and longtime student of Mao Tse-tung

Marion Had Being Conducting MW

Marion had followed many of the precepts of modern-day 3GW (of which MW and GW are variants). (See Table 6.1.) Among those precepts were little worry about unit functioning or backwards movement.

> He [the Swamp Fox] focused exclusively on the enemy, instead of terrain objectives. He possessed a maneuverist mindset, always maneuvering his highly mobile force into an advantageous position from which to attack his enemies. Marion knew when to attack and retreat to the swamps in

Maneuver Warrior	Attrition Warrior
Tries to Bypass or Demoralize Foe	Attempts to Kill Opponent
Depends on Surprise	Relies on Firepower
Focuses Outward on Enemy	Concentrates on Self
Harnesses Reconnaissance Pull	Relies on Command Push
Moves Through Undefended Gaps	Hits Enemy Concentrations
Exploits Breakthroughs	Stays on Line with Others
Captures Centers of Gravity	Seizes Hilltops
Employs Combined Arms	Uses Biggest Weapon
Prefers Non-Illuminated Offense	Assaults Under Illumination
Seeks High Tempo	Is Slow and Methodical
Defends Only as a Trap	Holds His Position Often
Sometimes Pulls Back	Never Retreats
Enjoys Delegated Authority	Is Centrally Controlled

Table 6.1: The "Maneuver Warfare" Difference

(Source: *Table 2.1 of "The Tiger's Way," by Poole.*)

> the face of a superior force. He leveraged his combat power by pitting his strengths against his enemies' weakness.[25]
> — U.S. Marine Corps Research Center

> Marion's *modus operandi* was to maneuver to a position of advantage, concentrate his force, execute a surprise attack on the enemy, disperse his force, and infiltrate [exfiltrate] back to his hideout.[26]
> — U.S. Marine Corps Research Center

The Swamp Fox may have the first American to generate the MW version of momentum. Instead of overpowering the British with firepower, he strung together tiny victories. That's how he created the illusion that Continental militia controlled the countryside. It so demoralized Cornwallis that he moved his army into coastal Virginia.

The Situation in Which Marion Found Himself

Things had been going badly for the Continentals in the South, when the Swamp Fox first began his operations. Almost single-handedly, he and his small force were then able to reverse the tide of war.

> Late in 1778 [facing a Northern stalemate], the British . . . turn their main effort to the south. Tory strength was greater in the Carolinas and Georgia, and the area was closer to the West Indies . . . Fleet. . . . A small British force . . . from Florida quickly overran thinly populated Georgia. . . . Congress sent General Benjamin Lincoln south to Charleston in December 1778 to command the Southern Army. . . . [I]n May, . . . the British commander . . . slipped around him to lay siege to Charleston. . . . The French and Americans mounted a direct assault on Savannah on October 9. . . . The British in strongly entrenched positions repulsed the attack in what was essentially a Bunker Hill in reverse, the French and Americans suffering staggering losses. . . .
>
> Meanwhile Clinton. . . pulled in his troops from outposts around New York, and prepared to move south against Charleston with a large part of his force. . . . [with] a co-or-

> dinated Army-Navy attack.... [O]n May 12, 1780, Lincoln surrendered his entire force of 5,466 men, the greatest disaster to befall the American cause during the war....
>
> After the capture of Charleston, Clinton returned to New York with about a third of his force, leaving General Cornwallis with 8,000 men to follow up the victory. Cornwallis established his main seaboard bases at Savannah, Beaufort, Charleston, and Georgetown, and in the interior extended his line of control along the Savannah River westward to Ninety-Six and northward to Camden and Rocky Mount. Cornwallis' force, however, was too small to police so large an area, even with the aid of the numerous Tories who took to the field. Though no organized Continental force remained in the Carolinas and Georgia, American guerrillas, led by Brig. Gens. Thomas Sumter and Andrew Pickens and Lt. Col. Francis Marion, began to harry British posts and lines of communications and to battle the bands of Tories.[27]
>
> — Center of Military History, *CMH_Pub_70-6-1*

Marion's Contributions to 1780

Marion's volunteers had started small, and then gradually gained confidence.

> <u>20 August</u>: Francis Marion . . . began his exploits by surprising [a] British guard escorting 150 American prisoners to Charleston at Nelson's Ferry (Great Savannah) crossing of Santee River, South Carolina, and killing or capturing 24 of [the] enemy as well as releasing patriots.
> <u>4 September</u>: General Francis Marion with group of about 50 partisans routed Tory force of 250 men at Blue Savannah, South Carolina.
> <u>29 September</u>: General Marion with 50 men defeated loyalist body of about [the] same size at Black Mingo Creek (20 miles southwest of Georgetown), South Carolina, with at least 20 of loyalists killed, wounded, or captured.
> <u>26 October</u>: Marion dispersed Tory force assembling near Tearcourt Swamp, South Carolina, in [a] surprise attack. Killed three and wounded 14, captured horses and other booty without loss.

12-13 December: Skirmish at Halfway Swamp, South Carolina, between Marion's men and British force escorting recruits to Hillsboro with inconclusive results.[28]

— Center of Military History, *CMH_Pub_70-6-1*

Marion's Exploits of 1781

By 1781, the Fox controlled the tempo of the war in the South and was starting to worry Cornwallis.

24 January: Henry Lee and Francis Marion with combined forces raided Georgetown, South Carolina, and captured British commander but, . . . withdrew when British garrison refused to come out and give battle.
12 May: British garrison at Fort Motte, South Carolina, surrendered to force under Henry Lee and Francis Marion.
17 July: Superior patriot numbers forced British garrison at Monck's Corner to retreat toward Charleston, and Marion's advance corps attacked withdrawing British at Quimby's Creek Bridge, South Carolina, on east bank of Cooper River about 15 miles from Charleston. After initial success in sharp action patriots had to withdraw, Marion's corps sustaining about 50 casualties and British about 70.
30 August: Patriots under Marion set up ambuscade at Parker's Ferry crossing of Edisto River, South Carolina, and claimed to have killed or wounded 100 of 200 dragoons sent out from Charleston (30 miles to east) to support local loyalist uprising.[29]

— Center of Military History, *CMH_Pub_70-6-1*

Summary of Marion's MW-Like Actions

Francis Marion had assailed British centers of gravity with the capture of the Georgetown commander and Fort Motte. He had surprised Redcoat contingents at Nelson's Ferry, Tearcourt Swamp, and Parker's Ferry. He had routed five times his number of Tories at Blue Savannah. And he had pulled back at Quimby's Creek bridge when things didn't go well. Those are all indicators of MW.

Such things so jeopardized Cornwallis' control of the South that he withdrew to the end of a Virginian peninsula. Only there could he be trapped by a massing of Allied forces.

Overview of the War

The War of Independence had shifted from going after enemy strengths (like in Boston or Canada) to enemy weaknesses (like individual outposts or resupply convoys in the South). This forced the Redcoats out of the countryside and back toward the coast. That's where a gradually tightening cordon at Yorktown would again mirror the Indian way.

According to Sun Tzu, "Strategy without tactics is the slowest route to victory."[30] Most tactics occur through small-unit maneuver. The Indian version—every individual fighting and withdrawing by himself—had been copied by the Swamp Fox.[31] Then, Alexander Hamilton used it during the nighttime capture of Readout 10 at Yorktown. There, every man was on his own after being told to unload his weapon for more surprise.[32]

This wonderful heritage of every tiny U.S. element—taking on its immediate adversary (however large)—would continue until the end of WWII. That may be why America had won almost every war to date. And, the subsequent string of ties and losses may be due to less initiative and self-reliance on the part of each frontline GI. As in football, winning squads must have opportunistic members.

Next Look at the Strongpoint Thread

During the U.S. War of Independence, most enemy bastions were frontally stormed during the day with fully erect troops. Those troops even remained upright while reloading. Only during the final stages of the conflict was the nighttime equivalent attempted without any shooting to increase the element of surprise.

7 U.S. Civil War

- Was the Indian way of fighting used much in the Civil War?
- Why not?

Fellow Americans had been forced to kill one another.

(Source: "Plenty of Fighting Today," by Keith Rocco, possible ©, from this url: www.nationalguard.mil/resources/photo_gallery/heritage/hires/Plenty_of_Fighing_Today.jpg.)

Indian Activity Before the War Between the States

Chief Dragging Canoe of the Cherokees had countered the white incursion until 1794. At the Battle of the Bluffs, 400 warriors briefly took over Fort Nashborough in Tennessee.[1]

At the Battle of Tippecanoe in 1811, the Shawnee Coalition tried to slow the settler flow into Illinois and Indiana, but was bested by local militia. This caused Chief Tecumseh to support the British in the War of 1812. At the Battle of Thames in Ontario, Tecumseh's Coalition again lost. Many Indians then abandoned the British cause.[2]

By 1814, pro-American Creeks were fighting those who resented the government. At the Battle of Horseshoe Bend in Alabama on March 27, U.S. militia helped the former against the latter.[3]

In the first Seminole War (1816-1818), tribesmen and runway slaves defended Spanish Florida against the U.S. Army. In the second Seminole War (1835-1842), the red men unsuccessfully fought to retain land in the Florida Everglades. After being outgunned and outnumbered in the third Seminole War (1855-1858), most agreed to move to reservations in Oklahoma.[4]

In 1832, Chief Black Hawk led around 1,000 Sauk and Fox Indians back to Illinois to reclaim their land. The ensuing Black Hawk War was a disaster for the Indians who were greatly outnumbered by the U.S. Army, militias, and other tribes.[5]

Comanches of Central and West Texas had resisted white settlers for years. Treaty negotiations turned sour when 35 Indians were killed and 30 more imprisoned. Accompanied by Mexican and Kiowan allies, the red men attacked and destroyed the port town of Linnville. But, Texan forces eventually defeated the tribe, forcing it away from the frontier.[6]

In 1864, 675 volunteer U.S. troops attacked a peaceful Cheyenne and Arapaho village along Sand Creek in Colorado. When its inhabitants attempted to flee, the soldiers killed 230 mostly women, children, and elderly.

Outnumbering the Foe Had Shaped U.S. Strategy

Among the Principles of War are "mass" and "maneuver." Native American problems were to continue out West throughout the Civil War (to include Red Cloud's uprising in the Wyoming territory).[7] Those Indians had been outnumbered in most battles. For this and other reasons, the U.S. Army's way of fighting had inexorably shifted toward mass by 1861—in the form of all-guns-blazing on-line sweeps. But, the British Army's burning of Washington in the War of 1812 had left an indelible scar on the American psyche. From that point onward, every enemy strongpoint was attacked with a three to one manpower advantage or greater. And, any increase in friendly casualties from so doing was tolerated.

Union forces then did poorly in their first outing at the First Battle of Bull Run (Manassas). Several Union generals spent more time preparing than fighting. And finally, President Lincoln put

Gen. Grant in charge. Often likened to a bulldog with his teeth in the seat of somebody's pants, Grant was quite determined. After getting beaten at the Wilderness, he immediately set off for Richmond.

All the while, Union contingents had been taking far more casualties than their Confederate counterparts at several places in the East. They had failed with 14 successive assaults to take Marye's Heights behind Fredericksburg (Dec. 62).[8] And, they were mostly outmaneuvered at Ball's Bluff (Oct. 61), Peninsula Campaign (Hampton Roads, Mar. 62), Seven Day's Battles (Mechanicsville, Jun. 62), Second Bull Run (Aug. 62), Chancellorsville (May 63), Spotsylvania Courthouse (May 64), the Wilderness (May 64), New Market (May 64), and Cold Harbor (June 64).[9]

Union contingents in the West had done only slightly better. They had taken more losses at Shiloh (Apr. 62), the Shenandoah Valley Campaign (Spring 62), and the Battle of Stones River (Jan. 63).[10] Though outmaneuvered the first day at Shiloh, the Yankees retook the field after being reinforced across the Tennessee River overnight.

From this combined experience, Union leaders had learned a dreadful secret. They could win most battles by simply throwing their edge in manpower and ordnance directly at the enemy concentrations. The extra cost in casualties was then easy to justify. The inherent dangers of war had been responsible, not any shortage of maneuver.

And, illogical as it may now seem, this same approach to warfare has followed U.S. forces into the modern age. With each succeeding conflict, the latest ordnance has seemingly removed any need to dodge counter-fire. And, other excuses—like Jane Fonda, *New York Times,* or Congress—help to explain the continuing loss of frontline personnel.

Only the Offensive Component of UW Was Missing in 1865

Throughout the War Between the States, both sides conducted the Revolutionary War version of GW.[11] There was a lot of hit-and-run tactics, ambushing, and raiding—some against rear-area "centers of gravity." But, the number of attacks involving short-range infiltration were few.

As the war progressed, both sides sometimes traveled before dawn to the jump-off places. The low light and ground fog helped to obscure their movements. But, when the sun went down, neither wanted much action. It was as if their mutual weariness with man's most obscene invention had invoked a truce.

The Confederates could have won the Battle of Shiloh, if Gen. Beauregard had ordered an all-out attack when the Federal Hornet's Nest collapsed about four o'clock in the afternoon. Culp's Hill was there for the taking on the first evening at Gettysburg. The Civil War's pivotal battle might have turned out differently if Gen. Ewell had honored Lee's wish to keep the fleeing Federals from entrenching on Cemetery Hill.[12]

It's difficult to understand why American soldiers see value in the stationary gloom of early morning, but not in the fleeting shadows of oncoming night. They'll risk an approach march in the fog before dawn, and then not crawl forward in full moonlight. Few armies show less interest in nature's ways of surprising an adversary.

Sizeable night attacks rarely occurred during the Civil War. There was a successful non-illuminated push by Union forces at Rappahannock Station in November of 1863.

> Spearheaded by the 6th Maine and the 5th Wisconsin, soldiers of the Federal VI Corps came out of the blackness, cheering as they charged the earthworks. Savage hand-to-hand fighting broke out; soldiers clubbed, stabbed and fired at close range along the line.[13]
>
> — *The Civil War: The Killing Ground,* by Jaynes

> The battle surged back and forth until a second charge led by youthful Colonel Emery Upton gave the Union regiments a decisive edge. Broken, the Confederates fled across a bridge or tried to swim to the south bank of the river, having lost four cannon, eight battleflags and 1,303 men. In the South, the Battle of Rappahannock Station was called a "mortifying disaster." The Union celebrated a victory that marked the first successful night attack of the War.[14]
>
> — *The Civil War: The Killing Ground,* by Jaynes

When Union forces first approached Richmond during the War

Between the States, they may have followed through on at least one offensive impulse before the sun came up. At least, that's what this excerpt suggests.

> In the predawn darkness of September 29 [in 1864], he (Grant) quietly slipped Birney...and Ord's XVII Corps back across the James. . . . Ord successfully stormed heavily armed but badly under-manned Fort Harrison on the Varina Road.[15]
>
> — *Battle Chronicles of the Civil War 1864,* by McPherson

And, in a valiant but ill-fated attempt to break through the deep layer of Union siege lines at Petersburg, Confederate forces attacked Fort Stedman in the pre-dawn hours with some stages resembling short-range infiltration.

> The ground in front of Fort Stedman was crisscrossed with picket trenches and covered ways and protected by what one Federal infantryman boasted "every . . . arrangement known to military science and art."
>
> Using all his innate guile and resourcefulness to overcome these formidable obstacles, Gordon devised one of the most methodical attack plans of the war. First, while it was still dark, special squads would open avenues through . . . [their own] Confederate defenses by quietly removing the obstructions that had been placed there. Then picked men (identified by strips of white cloth tied around their shoulders) would infiltrate forward to take out the advanced Union picket posts and open pathways through the Federal abatis. Behind this group would come fifty men with axes, who were to chop openings in the fraises protecting Stedman and its batteries.
>
> Right behind the ax men would be three storming parties of a hundred men apiece, whose job it was to rush and secure Fort Stedman and Batteries X, XI, and XII. After the storming parties had achieved their objective, three more groups of a hundred men each were to undertake a special mission . . . (seizing) several strongpoints well in the rear (that is, to the east) of Stedman....

> As the special-mission groups headed toward their objectives, the bulk of Gordon's infantry would cross over the no man's land to widen the breach by sweeping north and south along the captured trenches. Once the infantry had secured the lodgement, a cavalry division would ride through the pocket to the enemy's rear, cutting communications and raising general havoc.
>
> Private Henry London, in Grimes's Division, [on 25 March 1865 about 3:30 A.M.] never forgot how the storming party, "with unloaded muskets and a profound silence, leaped over our breast works, and dashed across the open space in front."...
>
> Within minutes, Fort Stedman, too [along with the other Batteries], fell....
>
> The first phase of Gordon's attack had secured nearly all of its goals. Fort Stedman and Batteries X and XI had been seized, along with the camps of the 29th and 57th Massachusetts regiments, and a hole nearly a thousand feet long had been opened in the Union line, with very small loss. First light was still forty-five minutes away.[16]
>
> — *The Last Citadel,* by Trudeau

Another Occasion of Short-Range Infiltration

During the 1860's, "infiltration" meant small groups of soldiers penetrating the enemy's line at weak points to later assemble. Such an effort preceded the main push on Jackson, Mississippi—a Confederate manufacturing center—in May 1863 as Gen. Grant and his Army of Tennessee were trying to take Vicksburg. About 40 elite soldiers were dispatched to weaken that city's defenses and create a diversion for the main onslaught. Their mission may have been to destroy its rail links to Vicksburg. That mission was soon facilitated by Gen. Johnston's abandonment of Mississippi's capital.[17] How many separate teams had tried to sneak through Confederate lines is not known.

Why Didn't More "Sapper Attacks" Occur?

By the 1860's. British resolve at Bunker Hill and high-noon

gunfights had started to shape America's battlefield ethics. Denying the other side a clear shot at every attacker seemed lacking in honor. Which societal and human pressures had led to such risky behavior aren't clear. Partially responsible may have been the dueling tradition, a distaste for bushwhacking, the brother-against-brother issue, or just misplaced bravado. American integrity was still quite high on most issues, so the extra loss of life from openly confronting enemy strength may have led to a mental block against its obvious drawbacks. But, for whatever reason, there was very little attempt to surprise Confederate defenders—no stealth, no deception, no diversions, no nothing. And every Union attack ended up being a "high-diddle-diddle-up-the-middle" evolution. Nor was there much composite maneuvering of separate elements within that Yankee assault line. Their job was to stay in formation.

Mostly to blame may have been Western culture's bad habit of trying to bring order to what is inherently chaotic. Instead of exploiting natural opportunities like the Asians and Indians had done, Union contingents tried to follow a time schedule. And the easiest way to do that is with frontal assaults behind a rolling artillery barrage.

But, extra friendly troops were dying, and the U.S. government would have to come up with adequate justification. At some point, its professed interest in each soldier's life would have to be reconciled with the lengthening casualty lists.

By the end of the 19th Century, this egregious battlefield oversight had found a scapegoat. Nothing could be done about the dangers of strongpoint storming. So, U.S. infantrymen would be so well protected from "short rounds."—through complicated fire support coordination—that little momentum was possible.

Every U.S. Soldier Now Belonged to a More Useful "Unit"

For the Union Army to take full advantage of its larger manpower pool, the combat "group" would have to take precedence over the individual fighter. Frontiersmen had trouble assimilating into a whole, so their self-sufficiency and initiative were thought to be more of a liability than an asset.[18] The well-led group was supposed to accomplish more than the collective effort of its members. It could certainly generate more firepower. But, there were inherent flaws in such reasoning. Though massed troops could shoot more in the assault, they also represented a bigger target. When all attacked at once in the daytime, no amount of suppressive fire could protect them from harm.

While this illusion of strength mostly applied to Union forces, it would go on to shape U.S. strategy for many years. Still, most extraordinary accomplishments came from small units. They sometimes acquired enough group chemistry to succeed with no leader at all (like with *gung ho* spirit). But, such things are the antithesis of top-down bureaucracy. So, neither short-range infiltration by a row of buddy teams, nor a succession of squad-sized assaults would become U.S. attack regimen. Not until late in WWI would any Western army try to fight that way, namely the German Stormtroopers.

How Courage Entered the Equation

U.S. leaders had sought infantrymen with enough gumption to remain upright in formation despite incoming shards of metal. While they still prepared their troops for close combat, such training normally involved "moves" so standardized as to be easily countered. And future GIs became no better at short-range fighting than the Redcoats had been during the Revolutionary War.

Ignoring enemy projectiles takes great courage and self-discipline, but so does a tiny foray into hostile territory. Why was one kind of bravery encouraged and the other not. Because, U.S. leaders were worried that unity might be compromised by too much small-unit initiative. And, the whole idea of semi-autonomous elements contributing separately posed great risk to the overall commander's career.

Military Honor Within the Mid-19th Century

While a "field of honor" has always existed in man's assorted dustups, it has taken on different dimensions with the various technological advances. During the Middle Ages in Europe, opposing generals sometimes fought each other to keep their armies from getting hurt. For a while, each contingent tried to outmaneuver the other, with the most agile declared winner without any actual bloodshed. Historically, the most honorable type of combat had been at close quarters. Then, standoff firepower seemed more "humane" to the armies rich enough to afford it. And, they didn't always have forward observers to protect innocent bystanders. By the time the War Between the States rolled around, battlefield honor had come down to giving the other side a chance to surrender.

Leadership in the War Between the States

While senior officers were often political appointees during the U.S. Civil War, many company-grade officers would be elected by their men. Such leaders tended to be very brave and combat savvy.

At first, most combatants were volunteer citizen soldiers. While not good at working together or following orders, they still had enough inner motivation to do well in battle.[19]

Degree of War Department Oversight

Created in August 1789, the U.S. War Department had a cabinet member in charge. It's job was to organize and maintain the U.S. Army under the command of the President.

During the American Civil War, that War Department took on new responsibilities. Through several new offices, it oversaw the recruiting, training, supply, medical care, transportation and pay of two million soldiers (in both the regular and temporary volunteer armies). And, a separate command structure handled all military operations.[20]

Secretary of War Stanton enjoyed immense power and didn't always do as directed. At first, he personally handled all walk-in requests. Stanton's "customary position was standing at a high, long desk, facing the principal entrance to the room, and open to all who had the right of audience, for he shunned every semblance of privacy in his office."[21]

The War Department oversaw the training of new regiments. Because their tiny subordinate elements seldom operated alone, they didn't require any tactics of their own. Civil War regiments largely learned how to fight in actual combat. Their training beforehand was limited to the manual of arms, a little target practice, and then large-unit procedures associated with basic maneuvers like skirmishing.[22]

Figure 7.1: Fort Wagner Being Unsuccessfully Stormed
(Source: Wikimedia Commons, public domain, image designator "Storming_Fort_Wagner_LCCN2012647346_(cropped).tif,")

By day's end, several hours had typically been devoted to drill. Fighting formations were generally linear and had to be heavily practiced to ensure all soldiers could function in the composite groupings. Individual companies didn't spent much time on target practice or firearms-related procedures. Most of their men were assumed to be frontier types already proficient with muzzle-loading rifles. Yet, those from the city had no such knowledge.[23]

> [T]hey [Civil War infantrymen] used tactics that had changed little since the time of the American Revolution or the age of Napoleon: infantry fought in closely knit formations of two ranks (or rows) of soldiers, each man in the rank standing side by side. This formation was first devised when the single-shot, muzzle loading musket became the normal weapon on the battlefield, the close ranks being a necessity

> because of the limitations of the musket. Yet, by 1861, new technology had made the old fashioned smoothbore musket nearly obsolete with the introduction of the rifle musket. By the time of the Gettysburg Campaign, the rifle musket made up the majority of infantry weapons in both the Union and Confederate armies though it took much longer for the tactics to change. Even with the advance of the rifle musket, the weapons were still muzzle loaders and officers believed that the old-fashioned drill formations were still useful to insure a massing of continuous firepower that the individual soldier could not sustain. The result of this slow change was a much higher than anticipated rate of casualties on the battlefield.[24]
>
> — *CivilWar.com*

How Often Civil War Soldiers Took Cover

Volley fire from muskets with rifled barrels was too effective to be endured for long. Yet, Union forces could not always obscure their approach march to an attack objective. At a considerable distance, they sometimes stood shoulder-to-shoulder while the equivalent of .50 caliber machinegun bullets passed through their four-abreast column.[25]

> The Springfield was a single-shot, muzzle-loading percussion musket [carried by both sides] with a 39 inch long rifled barrel, and could fire a .58 caliber bullet . . . [several hundred yards].[26]
>
> — *Civilwar.com*

Birth of an Unsettling Tradition?

As soon as Grant started to press the Confederates in earnest, the last stanza of a patriotic song took on new meaning for Union soldiers.

> As He died to make men holy, let us die to make men free.
> Chorus: Glory, glory, halleluja.[27]
>
> — "Battle Hymn of the Republic"

The Fredericksburg calamity had occurred in late 1862. When Union soldiers conducted another suicidal assault against Fort Wagner on Charleston harbor in July 1863, the inevitability of "battlefield glory" began to resonate among many Federal troops. (Look again at Figure 7.1.) From that point onward, U.S. infantrymen would have openly to assault prepared enemy positions.[28] And, within their ranks was a terrible realization—so doing would likely cost them their lives. As recently as 2021, a Vietnam-era Marine veteran has talked of he and his buddies still on a "Glory Road."[29] Needless to say, more tiny element maneuvering during history's biggest attacks would have led to fewer friendly casualties. At least, that was the collective conclusion of the Russians, Germans, Japanese, and Chinese right after WWII.[30]

The Ongoing Strongpoint Thread

During the War Between the States, Southern fortresses were often stormed from the front by fully erect Union troops after a preparatory artillery barrage. Toward the end of the war, huge explosive charges were sometimes emplaced by tunnel to create a gap in Confederate lines. But, very little short-range infiltration was ever attempted. Both sides had good scouts, but neither had squads specially trained to single-handedly force open a frontline breach.

8 Western Indian Wars

- Were the Plains Indians any good at night fighting?
- Why isn't that better known today?

All Plains Indians were proficient at close combat.

(Source: *Corel Gallery Clipart,* Totem Graphics, Man Miscellaneous, 28V018)

The False Rumor

It's still widely believed that indigenous residents of Western America tried to avoid all combat at night. As in police work, the truth will take a variety of sources.

Most tribes worried about entering the spirit world after dark. But, only the ritualistic Apaches may have totally spurned after-hours fighting. At least that's what a famous movie claims.[1] Then, an on-line encyclopedia pointed out that Geronimo and 250 followers fought all night after the Battle of Apache Pass.[2]

So, for all native Americans in the Western U.S., warlike activities were certainly possible during those periods of reduced visibility. That low-risk encounters were normally involved may have been for some other reason. The average Plains Indian routinely "counted coup" to establish his personal degree of courage. This was accomplished in one of two ways: (1) safely touching a foe with a bare hand or "coup stick"; or (2) stealing his horses.[3] Low illumination facilitated both. So, many a settler camp or army bivouac lost its remuda after the sun had set. And, other ways may have existed for an Indian to bolster his reputation.

Western Indians didn't like making war with their vision impaired, but would periodically do so. They used the cover of darkness to move into position for a dawn attack. And, such an attack was normally preceded by a thorough reconnaissance.

Some Specifics of This Nighttime Capability

According to the novelists and filmmakers, American Indians liked to attack as the sun was rising—to partially blind their adversaries. But, that's not what always happened. Between 1850 and 1890, in struggles for which time of day was recorded, the Indians attacked in broad daylight 110 times and at dawn only 10 times. During the same period, the U.S. Army mounted 221 daylight attacks and 44 at dawn. The American soldiers and Indians made only six nighttime assaults apiece. So, both sides preferred a daytime fight.[4]

Sacagawea, the Indian woman who acted as interpreter and guide for the Lewis and Clark expedition of 1804–06, is thought to have been a member of the Wind River branch of the Shoshone Tribe. After acquiring horses, Comanches then split off from the Wind River clan to move south into Texas.[5] There, they became quite active after dark. Chief Buffalo Hump's Comanches fought all night in the Great Raid of 1840 along the South Texas coast. And, some tribe conducted a predawn attack in June 1874 at a remote trading post on the Texas Panhandle known as Adobe Walls.[6]

Meanwhile, a settler company from Michigan had been traveling the Oregon Trail near Fort Hall in the summer of 1859 when attacked at night by assumed Shoshone.[7] And, Indians ran an "after-hours" raid in 1864 on the Kelly-Larimer wagon train in Wyoming.[8]

More About the Comanches

The Comanches were unique among the tribes. Not only did they halt Western expansion for 40 years, but actually pushed some settlements eastward. Panicked emigrants left the Texas prairies in scores. They abandoned their farms and ranches after many raids had been conducted under a "Comanche Moon."[9] The name Comanche is derived from a Ute word meaning "anyone who wants to fight me all the time."[10]

While trying to stop Comanche cattle raids, U.S. Army officer Mackenzie heard a band of Indians was camped along a riverbank 65 miles northeast of Amarillo. With 222 soldiers and nine Tonkawa scouts, he surprised the Indians and routed them. After the survivors fled on foot, he put their horses under guard. That night, the Comanches slipped back into his camp and took back their ponies.[11] That midnight raid would be Quanah Parker's first indication of how hard taking back his homeland would be. Thus began the Battle of Blanco Canyon—the opening salvo of a bloody war in the West Texas highlands that would last four years and end with the virtual annihilation of the Comanche nation.[12]

Was There Any Other Reason for Horse Stealing?

It was an old High Plains trick to take a white man's horse and leave him to die of thirst or starvation. Comanches had used it to lethal effect against the Spanish in the early 18th Century.[13] Defeating invaders without firing a shot is an advanced concept even today. At night, Comanches would stampede a foe's horses to let the harsh environment finish him off.[14]

Other Pony-Interested Tribes

Many Plains tribes, including the Cheyennes, took horses at night. Those Cheyenne warriors weren't interested in total victory, only occasional glory. Set-piece battles were avoided, and all battlefield initiatives by stealth, surprise, and maneuver. If the enemy became alerted, the Cheyenne braves withdrew to try again later.[15]

During one raid, a young Comanche took a Kiowa woman as his wife. The woman wanted to know if her folks were still alive. The Comanche told her to go see. Upon returning, she told her husband that Kiowas planned to raid the Comanche camp at midnight. Quanah Parker prepared the camp, making everything seem as normal as possible. When the Kiowas showed up, they were massacred.[16]

One night in 1861, the Chippewa and Sioux also got into a tussle. As one might guess, horses had been at the center of the dispute.

> Last Summer while . . . Grant was encamped with a party of hunters on Mouse River, twelve horses were stolen from him by a party of Yankton Sioux. . . . [On the 10th of June] a party of thirty-six Sioux warriors . . . arrived at St. Joseph with the stolen property, for the purpose of returning it. . . . about two o'clock in the afternoon; they immediately crossed the river and proceeded to the residence of Gov. Wilke. . . . [A] large party of Chippewas [subsequently] fired on the Sioux while they were . . . entering Gov. Wilke's house. . . . From this time until midnight, a constant fire was kept up between the Indians. Six Chippewas, three Sioux, and two Assiniboines were killed during the struggle. . . .
>
> The house occupied by the Sioux is about two hundred feet from the river. The Chippewas surrounded it, and, to make their escape, the Sioux were compelled to run that distance and wade the river to the south bank, in the face of a constant fire from their enemies. This they did, aided by the darkness of the night, with the loss of one warrior, who was found dead by the Chippewas in the morning, on the south bank of the stream. The Sioux left behind them thirty-two horses, (in addition to the twelve stolen ones,) and the dead bodies of three of their warriors.[17]
>
> — *St. Paul Pioneer* (MN), 12 July 1861

Most Native Americans Had Become Used to the Dark

Because the Plains tribes had historically amused themselves with night raids to count coup, their young men became good at low-light maneuvers. One *ninjutsu* authority claims all Indians would attack at night to "take horses or slaves and cut throats." Because all three involved reduced risk, the third may have qualified for counting coup. That same researcher claims preventing their foes from doing likewise made most Indians expert at "stealthily moving through the brush." And, this expertise likely extended to the entire nighttime portfolio: scouting, stalking, shadowing, and infiltration. He goes on to say that such skills were far superior to those of the settlers (including previous soldiers).[18] Recorded history supports this little known fact.

> Whatever surprise Mackenzie had hoped for [in attacking the Indians] was gone. On the third day the Tonkawa scouts realized they were being shadowed by a group of four Comanche warriors, who had been watching their every move, presumably including what must have seemed to them the comical blunders of the night march. The Tonks gave chase, but "the hostiles being better mounted soon distanced their pursuers and vanished into the hills." This was not surprising. In two hundred years of enmity, the Tonkawas had never been close to matching the horsemanship of the Comanches. They always lost. The result was that, while the [U.S.] cavalrymen and dragoons had no idea where the Comanches were camped, Quanah knew precisely what Mackenzie was doing and where he was.[19]
>
> — *National Public Radio,* May 2011

The Comanches were all about stealth. They made cold camp as they moved, so there would be no signature fire. They swam their horses through frozen rivers that appeared uncrossable. And, they snuck up on quarries at night.[20]

Raiding Indians Preferred a Well-Lit Enemy Camp

Comanches attacked in full moonlight, and liked a foe with many fires. Then, the horse stealing served more than one purpose. The ensuing mayhem lowered the raiders' degree of risk.

> The next night Mackenzie compounded the error by allowing the men the indulgence of campfires. . . .

> At around midnight, the regiment was awakened by a succession of unearthly, high-pitched yells. Those were followed by shots, and more yells, and suddenly the camp was alive with Comanches riding at full gallop. Exactly what the Indians were doing was soon apparent. Mingled with the screams and gunshots and general mayhem of the camp was another sound, only barely audible at first, then rising quickly to something like rolling thunder. The men quickly realized, to their horror, that it was the sound of stampeding horses. Their horses. . . .
>
> When it was all over, the soldiers discovered that Quanah and his warriors had made off with seventy of their best horses and mules.[21]
>
> — *National Public Radio,* May 2011

Had Someone Drunk Too Much of Their Own Kool-Aid?

The Western Indian Wars may provide the first example of the U.S. military avoiding embarrassment through an unsubstantiated rumor Had the U.S. Army's lack of night-fighting ability occurred because its new foe "didn't like the dark?" U.S. forces only attacked at night a few times, and never as tiny elements. Nor had many Yank soldiers been shown how individually to fight in low light. The U.S. Army may not have started that mid-19th Century rumor, but it did little to stop it. And, it wouldn't develop many of its own night-fighting skills for another 100 years.[22]

Today's unsubstantiated rumor is that night-vision equipment has reached such an advanced stage that no adversary dares to attack Americans at night. Yet, the same gadgets have yet to peer through heavy rain, fog, thick bushes, or tiny folds in the ground.[23] And, a distant flare or bright moonlight renders them useless. Continually refining ordnance makes someone a nice profit. With more standoff armaments has now come the insinuation that "boots on the ground" and short-range infantry excellence are no longer needed. Recent history begs to differ.

The Modern-Day Sequel

On a fully moonlit night in early 1968, NVA soldiers managed to get within grenade range of U.S. Marines defending a cone-shaped hill at the entrance of a dangerous valley in Vietnam. They did so by moving so slowly between the bushes that no motion was discernible in the dim light.[24] Might there be a correlation between that moon and the Comanche Moon? Before any ground assault, native Americans had stalked defenders by keeping a bush between them and their quarry. Rushing range and grenade range are one and the same in the Finnish Army.[25] So, the NVA soldiers had closely approximated a Comanche attack. They hadn't learned it from the Comanches, just depended more upon frontline observations than headquarters imagination.

The night-combat variation between an Asian authoritarian army and Western volunteer army derives from vastly different societal customs. While the Eastern force grows stronger at street level through a heritage of bottom-up decision making, the lowest echelons of the Western entity atrophy through overcontrol.

How Battlefield Integrity Had Fared in the Old West

Within most, if not all, of the Western Indian tribes, not telling the truth was greatly frowned upon. "Don't lie" has been noted as a native American motto by many Western books and movies.[26] Their distant relatives, the East Asians, saw dishonor in claiming to have won a battle they didn't. That's why Mao and the North Vietnamese allowed their troops to monitor all after-action reports for exaggeration.[27]

But, in the late 19th Century, it was U.S. authorities who broke most of the peace treaties. Both Cochise and Geronimo had been lied to,[28] and then Osceola captured under a flag of truce. So, the two cultural groupings failed to agree on what constitutes battlefield integrity. From that point forward, U.S. forces would win virtually all of its enemy engagements, however large—something quite impossible under the rules of probability. And, who actually took the fewest casualties during those engagements soon became suspect. In the old days, Manifest Destiny may have periodically altered the truth. But, as any good Christian knows, the tiniest deviation can easily multiply. No U.S. overseas conflict has been decisively won since WWII, yet the U.S. military still claims to be the best in the world at all things. Might stretching the truth have become so endemic that professed capabilities are no longer valid?

The Central Paradox

Within America, voting more greatly affects local decisions, than those at the state and national level. That's why the military capabilities of an autocratic Eastern nation may surpass those of a free Western nation. Sophisticated small-unit tactics more easily exist in the former, and true momentum. How can such a shocking paradox be possible? Only in the soon-to-be disbanded Carlson's Raiders of WWII would Maoist *Gung Ho* sessions serve the same purpose as *Kiem Thao* sessions do in present-day Vietnam.[29] Through a similar system, North Korea could also develop more battlefield flexibility than freedom-loving (and law-abiding) America.

Legitimate Conclusions So Far

It should now be apparent that all indigenous Westerners were good night fighters—the equal of any modern-day East Asian soldier. It's further likely that most Plains Indian night attacks involved a diversion or some other deception to limit the number of friendly casualties. Why hasn't the U.S. military followed suit? Because, from the earliest days of the nation, the foe's tactical expertise has been understated to keep firepower-reliant U.S. forces from having to admit any small-unit deficiencies.

From the Comanche reliance on ponies, the reader should not conclude that mounted infantrymen are as powerful as the dismounted kind. Jeeps aren't as mobile as horses, and the Comanche were just as agile on foot. In the modern day, mechanized infantrymen only do well in the open desert or prairie.

Warfighting in Other Parts of the Country

No study of the Western Indians would be complete without some mention of the Nez Perce. On a fighting retreat of 1,170 miles across four states and several mountain ranges, they did battle with over 2,000 U.S. soldiers.[30] With never more than 250 warriors, this tribe was to thwart their pursuers in no fewer than 18 encounters. More than 100 U.S. soldiers and 100 Nez Perce (including women and children) would be killed.[31]

According to the U.S. Forest Service, the Nez Perce fought in over 20 engagements and five major battles.[32] Another source has 145 warriors and 500 noncombatants contesting the U.S. Army in four major battles. Time and again, they outmaneuvered the soldiers while struggling to reach the Canadian border.[33]

Known Specifics of the Running Battle

The Nez Perce first got the best of Perry at the Battle of White Bird Canyon.[34] The next big engagement was on 9 August 1877. Fleeing eastward, they stopped for the night at Big Hole Basin.

> The third battle of the Nez Perce War occurred on this day in 1877. . . . [T]he Nez Perce made camp in the Big Hole Basin in present-day Montana. At 3:30 A.M., Colonel John Gibbon attacked the sleeping Indians with a force of 183 men. Raking the Indian lodges with withering rifle fire, the soldiers initially seemed to be victorious. The Nez Perce, however, soon counterattacked from concealed positions in the surrounding hills. After four days of sporadic fighting, the Nez Perce withdrew.[35]
>
> — *History.com,* n.d.

Almost certainly, the Nez Perce were developing sophisticated Escape and Evasion (E&E) skills. Among them would have been nighttime reconnaissance, counter-tracking, and an appreciation for microterrain that would equal any wild animal.

> Clark Fork Canyon, Wyoming. The war against the Nez Perce could have ended right here. After months of chasing the Nez Perce from behind, General Howard finally got some help from troops approaching from the east.
>
> Howard was sure that his troops in Yellowstone -- and the men of the Seventh Cavalry, led by Colonel Sturgis on the east side of the Absaroka Mountains -- would close the snare. "Seems hardly possible that they can escape this time," he wrote in a message.
>
> But with ingenuity and determination and luck . . . , the tribe slipped the cavalry's trap in an escape that has become legendary. . . .

> The Nez Perce did indeed move south toward the Shoshone River, but not very far. Instead, they found an open spot and cleverly concealed their trail. They milled their horses around in every direction, creating confusing tracks that seemed to show the Nez Perce scattering. Then, instead of traveling in the direction they had been headed -- out of the basin and across an open plain -- the Nez Perce turned to the north, traveling along a steep, timbered mountainside for several miles.
>
> They then took a steep drainage to the mouth of Clark Fork Canyon, not far from where Sturgis' troops had camped the day before. The trek through the narrow canyon was no doubt difficult, especially with 700 people and their possessions. Howard later marveled at the tribe's passage through what he called "a strange canyon, where rocks on each side came so near together that two horses abreast could hardly pass."[36]
>
> — U.S. Dept. of Agriculture Forest Service

Like Mao's 8th Route Army in its 6000-mile retreat of 1934 and 1935, those Nez Perce found ways to fight effectively while moving backwards under fire. Their "holding-action traps" would become part of "retreat combat," a 3GW variant.

Further Revelations

In the late 19th Century, individualism was a concept quite alien to Western military thought.[37] Any deviation from a group effort was discouraged among the rank and file. But, the American Indians used it to their advantage.

> For the native [American] warrior, individualism was as important as cooperation. A cult of bravery dictated that he must take great risks in order to prove his manhood and elevate his standing in the tribe. . . . Trophies were more important than . . . territorial gains in war. . . . Horses, women, scalps and he like were more prized than taking a foe's life. . . . [G]iven the confusion and noise, and often in darkness, leaders maintained very little control over their men in a fight. . . . [W]hile an Indian attack began as

> a closely planned and coordinated affair, at the first war whoop it disintegrated into as many individual battles as there were warriors.[38]
>
> — *Frontier Skills,* by Davis

By often operating alone, each warrior developed enough personal technique to survive close combat. He never allowed his foe a long enough sight picture. "Whatever the weapon in his hand, the fighting native stayed constantly in motion, up and down in the brush, darting and dodging back and forth, to make himself as elusive a target as possible." After counting coup as many times as was prudent, he melted back into the shadows and headed for home.[39] So doing turned every large-unit attack into a raid.

But, on the way to an objective, Indians would travel as a group. The war party stayed low, moved when possible through gullies and ravines, avoiding ridgelines, and stayed off skylines. On dusty ground, its members afixed animal tails to their moccasins to prevent any sign of passage.[40]

The Seminole Clinic in Florida

In the First Seminole War of 1817-1818, forces under Gen. Andrew Jackson attacked key locations in Spanish Florida, and pushed the Seminole tribe farther south.[41] But, the Seminoles had shown an unexpected degree of perseverance.

> The Treaty of Payne's Landing, signed by a small number of Seminoles in May 1832, required Indians to give up their Florida lands within three years and move west. When the U.S. Army arrived in 1835 to enforce the treaty, the Indians were ready for war. . . .
>
> The campaigns of the Second Seminole War were an outstanding demonstration of guerrilla warfare by the Seminole. The Micos—Jumper, Alligator, Micanopy and Osceola—leading less than 3,000 warriors, were pitted against four U.S. generals and more than 30,000 troops.
>
> The Second Seminole War of 1835-1842 . . . was the fiercest war waged by the U.S. government against American Indians. The United States spent more than $20 million fighting the Seminoles. The war left more than 1,500

> soldiers . . . dead. And the . . . duplicity of the U.S. government's [negotiations] . . . marred Indian-white relations throughout the country for future generations. . . .
>
> . . . Osceola was captured and imprisoned when he met with U.S. troops who had called for a truce and claimed to want to talk peace. . . .[42]
>
> — Florida Department of State

The Seminoles had mostly fought while outnumbered. Now remembered as GW, this sort of strife had more to do with tiny teams swarming. And, all Seminole braves freely functioned at night.[43] To control them, U.S. units captured their leaders through a breach of etiquette and shipped everyone to Oklahoma. After the Plains Indians had been lied to, they were shipped to Florida.

The Most Common Form of U.S. Army Attack

The best teacher in war is the foe. While U.S. forces enjoyed good teamwork, they soon realized why large units had trouble totally surprising an opponent. To compensate, they tried to surround the Indians or catch them with their backs to a river or bluff. Then, they preferred standoff rifle fire to a ground assault.[44]

Roots of the Uncanny Similarity Between U.S. Foes

Native Americans had fought in the late 19th Century much like East Asians operated in the 20th. Both had tried to mesh any battlefield push with a natural occurring event, like waiting for a cloud to cover the moon before assaulting. Within the Navajo religion is the concept of *Hozho*. Loosely translated, it means inner balance and harmony. To be "in *Hozho*" is to be at one with, and a part of, the surrounding environment. But, U.S. infantry outfits just do as told as quickly as possible. More schedule-oriented and impulsive, they regularly try to go where there is no natural opening.

Today's Asian commander also pays close attention to details. A student of every aspect of the combat situation, he seldom asks subordinates to exceed their capabilities. His thought processes have been shaped by Taoism—an ancient Oriental philosophy of self-realization through patience, simplicity, and harmony with nature.[45]

Cooperating with innate processes is an Eastern tradition. Of utmost importance is remaining unseen in battle. Grandmaster Hatsumi calls *ninjutsu* the art of stealth, and *ninpo* the way of invisibility.[46] To practice them, one must learn the laws of nature—not only of vegetation and microterrain, but also of light, shadow, and sound. He also needs self-discipline and to pay close attention to what the enemy is doing. The *ninja* totally immerses himself in the mission.[47] According to Hatsumi, that *ninja* merges into his background and is acutely responsive to changing circumstances.[48] He creates no form or shape that is out of place.[49] He doesn't overreact to what is happening but flows with it.[50] And, his techniques flow as well.[51] He has great forbearance and considers no obstacle insurmountable. In effect, he just keeps going.[52] He tries to bypass the foe by studying his weaknesses.[53] While also aware of his own shortcomings, he conceals them. He can do things that "big-picture thinkers" deem impossible. Most distressing is the ease with which a *ninja* can capture or kill a Western scout. The *ninja* has only to hide behind a big trail-side tree. As the scout passes the tree, the *ninja* slides along its opposite side to get behind his quarry.[54]

Another Close Comparison

The constant motion of native Americans in close contact with U.S. soldiers is like what Giap described during the First Indochina War. "It is the fight between tiger and elephant. If the tiger stands his ground to fight, the elephant will crush him with its weight."[55] Giap's emulation of wild animals is perfectly logical for anyone who respects nature. Through survival of the fittest, many animals become quite good at moving around undetected. A closer attention to one's surroundings applies to all Eastern soldiers.

> In World War II, as in preceding wars, the Russian soldier demonstrated that he was closer to nature than his West European counterpart. This was hardly surprising since most of the Russian soldiers were born and raised far from big cities. . . . The Russian was able to move without a sound and orient himself in the darkness. . . . [He] performed

> particularly well as a night observer. Stern discipline and self-constraint enabled him to lie motionless for hours and observe the German troops at close range without being detected. . . . Infiltration by small detachments, as well as by larger units up to an entire division, was probably the most effective Russian method of night combat. . . . Time and again their troops slipped through a lightly held sector during the night and were securely behind the German front by the next morning.[56]
>
> — U.S. Army, *DA Pamphlet 20-236*

The Asian's more thorough understanding of nature also helps him to outsmart Western technology.[57] Not only will rain, fog, bushes, ground ripples, and a bright moon thwart the latest target acquisition device, but so will smoke, flares, and a little digging.

Of Further Importance

Relying on each soldier's contribution—like what American Indians did for personal reputation—now mostly occurs in the Eastern World. Despite U.S. propaganda to the contrary, "cannon fodder" is what a Western infantryman normally feels like. Mao's viewpoint was this: "It is people, not things, that are decisive in war."[58]

> Whether in guerrilla . . . or limited regular warfare, . . . it [armed Communist struggle] is fully capable of . . . getting the better of a modern ["high-tech"] army. . . . This is [through] the development of the . . . military art, the main content of which is to rely chiefly on [the] man, on his patriotism and . . . [individual] spirit.[59]
>
> — Gen. Vo Nguyen Giap

> Japanese strength . . . [also] lay in small units and the epitome of Japanese doctrine was embodied in small-unit tactics. Night attacks . . . and the willingness to engage in hand-to-hand combat were the hallmarks of the Japanese infantryman.[60]
>
> — U.S. Army, *Leavenworth Papers No. 2*

The nonviolence of counting coup remains quite advanced for

military doctrine. According to Pope John Paul II, the least killing of enemy soldiers leads to the most effective and moral forms of modern combat.[61] Of note, Sun Tzu had said something similar 2500 years ago.

> Only when the enemy could not be overcome by these [political] means was there recourse to armed force, which was applied so the victory was gained: (a) in the shortest possible time; (b) at the least possible cost in lives and effort; (c) with infliction on the enemy of the fewest possible casualties.[62]
> — *The Art of War,* by Sun Tzu

A Civil War Habit That Would Be Hard to Break

In the Overland Campaign of 1864, Grant had grabbed the Confederates "by the seat of their pants," and never let go. Such a strategy became standard U.S. operating procedure after the war.

> Grant knew he had the numerical advantage in troop strength and wasn't afraid to sustain high casualties in the short term in the hope that it would save lives in the long term by hastening an end to the war.[63]
> — *History Channel*

American troops in the Far West then had great difficulty occupying an Indian camp by complete surprise. Most attacks were through standoff fire. They were followed by a hot pursuit, in which deception, surprise, and tactics played little part.

> Many [native Americans] chose to fight, and over the course of some twenty-five years the struggle ranged over the plains, mountains, and deserts of the American West, a guerrilla war characterized by skirmishes, pursuits, massacres, raids. . . .
>
> [T]o a large degree the [U.S. Army] officers of the Indian wars were products of the Civil War. Many brought outstanding records to the frontier, but this was a new conflict against an unorthodox enemy. Those who approached their new opponent with respect and learned his ways became the best Indian fighters. . . . Some who had little respect for the

"savages" and placed too much store in Civil War methods and achievements paid the penalty on the battlefield.[64]
— Office of the Chief of Military History,
U.S. Army

Some Small-Unit Learning Had Undoubtedly Occurred

One doesn't fight American Indians for over a hundred years without learning something. The need for Indian scouts had been realized and how effectively to operate a perimeter defense. But, offensive technique improvements are difficult to identify in the surviving body of knowledge.

The Strongpoint Thread Revisited

During the Western Indian wars, enemy concentrations were more often attacked through long-range fire, than direct infantry assault. Large-unit assaults couldn't generate enough surprise to catch the wily Indians off guard. Unfortunately, U.S. tacticians interpreted this to mean that a big enough base of fire made a successful infantry assault possible overseas. Not without excess casualties, it didn't.

9 THE SPANISH-AMERICAN WAR

- Which errors began to appear in the U.S. assault method?
- How did they involve the latest technologies?

Assault by Teddy Roosevelt and his 1st Volunteer Cavalry.

(Source: "The War in Cuba", Wikimedia Commons, public domain, image designator "The_Rough_Riders_by_Mort_Kunstler.jpg.")

Little Had Changed Tactically Since 1865

When U.S. troops went Cuba to fight the Spanish at the end of the 19th Century, their attack methods were about the same as during the Civil War. Several units would try to take Santiago.

On July 1st, 1898 the largest and bloodiest battles of the Cuban campaign were fought for possession of Santiago. Santiago itself sat among a collection of hills and mountains bristling with Spanish fortifications. Over the course of five years the Spanish had built up a system of defensive trenches and blockhouses among the hills that had few rivals. Their defensive line extended for three miles, starting with San Juan Hill in the South and proceeding north in an unbroken line of fortifications terminating at the sea. The 71st [New York Volunteer Infantry] and the two regular regiments toiled along the small thin mountain road that led to San Juan Hill having been proceeded by the Rough Riders. The firing along the road was severe and the Spanish troops concealed in the heavy thickets used a smokeless powder, which made their positions very difficult to discern. Their sharpshooters were concealed in many of the tall coconut trees along the path and their artillery [modern rapid-fire breech-loading cannon with smokeless powder] from the hills poured fire onto the road, which they knew the U.S. troops had to use.

The fighting for possession of San Juan Hill itself degenerated into a series of successive charges by individual regiments, battalions, and companies. Among the units to reach the top first were the 6th, 13th, 16th, and 24th Regular Infantry Regiments along with Company F of the third battalion of the 71st Regiment. Much of the rest of the 71st was also heavily involved in the fighting . . . on the right side of the hill. Shortly after the fall of San Juan hill the Spanish launched a heavy counterattack on the captured trenches at night, but they were badly mauled and forced to retire after losing an estimated 3,000 soldiers. The 71st remained in the trenches for the remainder of the fighting at Santiago, conducting the siege duty.

The Spanish garrison at Santiago surrendered on the 14th of July, having been suffering from famine for over two weeks, while simultaneously watching the U.S. army position it's imposing siege artillery. The Spanish surrendered some 24,000 troops at Santiago; all of whom were to be transported back to Spain.[1]

— New York State Military Museum

The Fight for El Caney

At the very edge of Santiago's defense line lay the town of El Caney. (See Map 9.1.) As U.S. forces approached it from the east,

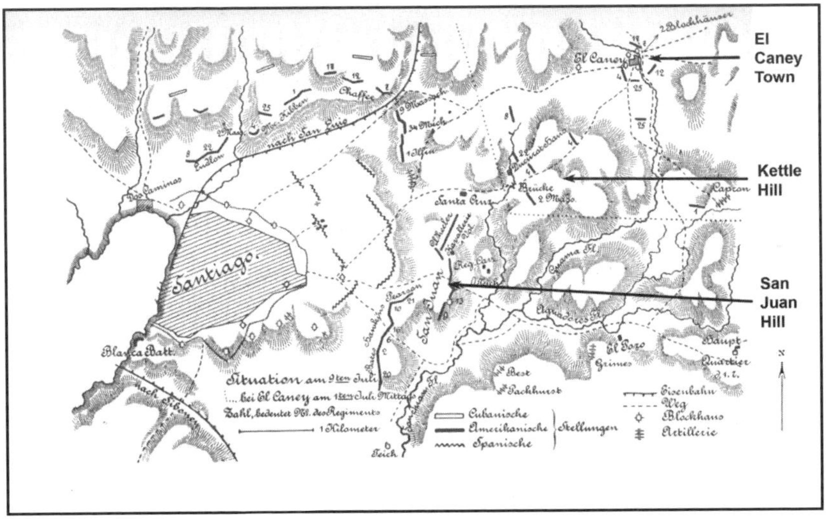

Map 9.1: Outer Bastions of the Santiago Defense Belt
(Source: Wikimedia Commons, public domain, image designator "177_of_'Der_Krieg_um_Cuba_im_Sommer . . . Mit_zahlreichen_Abbildungen'_(11253877006).jpg.")

they ran into long-range fire from a "small arm" soon to become famous. After deploying on line in the thick brush, the Yanks had trouble maintaining contact with one another. Then, they encountered a series of sniper-infested kill zones.

> On July 1, 1898, General Ludlow's First Brigade slowly approached El Caney. . . . "The 22nd Infantry led the advance along a trail overgrown with brush and vines until we reached the main Santiago-El Caney road near the Ducoureaud House. The Second Battalion was then deployed and skirmished northward through the jungle to check if there were any paths over which the Spaniards in the town might escape to Santiago. The First Battalion continued along the main road toward El Caney. About a thousand yards from the city, Company A, our advance guard, came under sharp Mauser fire. The battalion deployed rapidly east and west of the road. LTC Patterson was severely wounded by

the sudden enemy attack, and Major Van Horne took over command of the Regiment. . . . For more than a half-hour we cut through undergrowth and tangled vegetation, which was so dense that the battalion could not see more than ten feet to the front. In order to keep the line in order, men were continually forced to call out their position to skirmishers on either side of them. . . . "After three hours of battle, the First Battalion had crept to within seven hundred yards of the enemy positions - suffering heavy losses in the process.

Meanwhile at El Caney, the Second Battalion of the 22nd Infantry, under the command of Captain B. C. Lockwood, was located to the extreme left of General Ludlow's line. The battalion received orders to hack its way through the jungle for half a mile, then swing east until they came to the Cuabietas road. This was to be their assigned position to block any enemy retreat from the town.

As Lockwood's troops cut their way through the heavy underbrush, they came under fire from Spanish snipers. The advance was so difficult that it was impossible to keep the formation together. . . .

. . . "The Spanish use of barbed wire proved to be very effective in stopping our advance. Wires were stretched near the ground to trip our men when they would run from one location to another. A few yards beyond, the Spaniards had constructed tall fences—strung with many lines of wire. These defenses were laid in cultivated valleys and other open spaces near the entrenchment's. Enemy snipers were posted in the treetops around the clearings. Every fence compelled a momentary halt in our progress—and during those moments, we were exposed to pitiless fire. From noon until one o'clock there was a lull in the fighting, and men with wire cutters moved forward. Taking advantage of bushes and dips in the ground, they crawled to within two hundred yards of the trenches and cut many of the wires. They returned to our lines safely. Owing to the great extent of the front occupied by the Regiment, General Ludlow moved the Second Battalion still farther to the west, and advanced the First Battalion to within five hundred yards of the enemy. The 8th Infantry was positioned between the First and Second Battalions."

About one o'clock, the Spaniards increased their firing with renewed vigor. The Americans were forced to hug the ground at the edge of the clearings, and strain their eyes for moments when they could catch a glimpse of the enemy. The fierce heat from the sun was intolerable - and water was almost impossible to obtain.

. . . The blockhouse and trenches at the south end of town were protected by tall trees. Not only did they conceal the enemy's movements, but snipers stationed in the trees became a major problem. The [foe] sharpshooters could not be seen, however, their fire was devastatingly accurate."

Because of the two-mile range of the high-powered Mausers, the American artillery was useless at close distances against the Spanish defenses. The most frustrating aspect about the failure of artillery support was the fact that the gun batteries were unable to cover the infantry assaults on the San Juan heights and El Caney adequately. . . .

At the same time that Chaffee's troops were assaulting the hilltop, Ludlow's brigade—led by the 22nd Infantry Regiment—rushed from their jungle cover and sprinted across the fire-swept clearing. Yard by yard, trench by trench, and blockhouse by blockhouse, the brigade fought its way into the town. The Spanish garrison was besieged on all sides. The enemy became confused and disoriented. They searched for a route to withdraw their forces to the safety of Santiago, but there was only one path still open—the Cuabietas road.

. . . Very few, if any, of the enemy escaped Lockwood's ambush. More than two hundred dead and wounded were counted along the intended retreat route.[2]

— Spanish-American War website

Overall Effect of the Spanish Defense

Behind the sniper-infested kill zones lay heavily entrenched hills, often capped by a big pillbox. The Americans wasted little time in attacking them.

After the Battle of Las Guasimas in Cuba, Major General William Shafter planned to take Santiago de Cuba, the island's second largest city. Reports of Spanish reinforcements on route to the city caused him to accelerate his plans. He ordered head-on assaults against three hilltop fortified positions that made up the city's outer defenses.[3]

– *Encyclopedia Britannica*

In Shafter's plan, the 2nd Division was to take El Caney, then swing south to join in a coordinated attack at 10:00 AM with the 1st Division, led by Brigadier General Jacob Kent, and the dismounted Cavalry Division, led by Brigadier General Samuel Sumner, against the two hills that sat on the San Juan ridgeline: San Juan Hill and Kettle Hill. Entrenchments, block houses, barbed wire, and several cannon protected the Spanish defenders.[4]

– *Encyclopedia Britannica*

Shafter's plan quickly fell apart. The march to attack positions was delayed, and unit deployment was confused by the narrow, crowded trail and enemy fire. At 8:00 AM, 1 July, artillery began firing on the Spanish positions, then ceased to avoid counterbattery fire. Although it was well after noon before all the units were on line, the 2nd Division was still heavily engaged at El Caney. Finally, at 1:00 PM, and under galling Spanish fire, an aide of Shafter's gave approval to attack. The Cavalry Division's two brigades, led by the 1st Volunteer Cavalry under Colonel Theodore Roosevelt, charged and captured Kettle Hill, though suffering heavy casualties. Losses: U.S., some 205 dead, 1,200 wounded; Spanish, some 215 dead, 376 wounded.[5]

– *Encyclopedia Britannica*

At the Base of Kettle Hill Was the San Juan River

At "Bloody Ford" on the San Juan River, one outfit about to assault Kettle Hill lost 400 killed and wounded.

The two divisions would head on to Camino Real. This was a muddy narrow road so the men had to march two abreast rather than columns of four. The front of the column came under heavy fire from the Spanish within a mile of the San Juan River. Lieutenant Colonel Roosevelt was ordered to branch to the right and head upstream while waiting for further orders. Lieutenant John Pershing along with his unit the 10th Cavalry (Negro) regiment was just reaching the ford. As units moved toward the ford, it was slippery and muddy due to the blood of the dead and wounded soldiers. Over 400 men were killed and wounded at the ford (renamed 'the Bloody Ford'). . . . The Spanish continued to fire on them while they were in the river. At 1305, the volunteers received the orders to advance in the assault on the hill but the regulars didn't get authorization. Roosevelt told the regular cavalryman to rush the entrenchments but they wouldn't, so Roosevelt and his unit passed them and they followed. Lieutenant Parker and his four Gatlings located in front of the Bloody Ford opened up heavy suppressive fire on the Spanish positions around the blockhouse. The American Army realized the Spanish dug their trenches on the topographical crest of the Heights, which shielded the Americans from the bullets. The Spanish were fleeing back toward Santiago. Roosevelt, the Rough Riders, 9th and 10th Cavalry advanced to the northern spur of San Juan Heights [Kettle Hill]. The San Juan battle had 125 casualties and collectively 1,000 men wounded. The Spanish had 215 casualties and 375 wounded . . . [who] became prisoners of war.[6]

— U.S. Army research paper

Pershing's Part in the Battle

During the next big encounter, the future head of all U.S. forces in Europe during WWI played an important role.

About 8:00 on the morning of 1 July came the crash of artillery, first American, followed by Spanish. For forty-five minutes the duel continued with the Americans getting the worst of it. Their black powder guns poured smoke, revealing their positions, while the Spanish guns, using smokeless powder, remained hidden. Near Pershing, a Hotchkiss gun exploded, wounding two troopers. The frightened Cuban insurgents who were with Pershing fled.

As the barrage subsided, the Americans started down the ridge and moved forward along a jungle path. Lt. Col. Theodore A. Baldwin, commanding the 10th, ordered Pershing to act as a guide for the regiment, making sure it found its objectives and kept an orderly advance. The task was difficult; scattered artillery and rifle fire rained down as the men mixed with elements of the 71st New York Infantry along clogged roads inadequate for such large numbers. Pershing could do little but sit on his horse and shout orders to the men. Then to make matters worse, an observation balloon was sent up above the advancing column, drawing fire and revealing the American route of approach. The Spanish soon concentrated their fire on the area around the balloon. . . .

Enemy fire intensified, and panic ensued as men fell everywhere. Eventually, by continually running back into the jungle, finding lost groups, and guiding them forward, Pershing managed to get the 10th across the river. During the action he was continually exposed to enemy fire. . . .

As the men of the division waited at the edge of a wooded area below the two American objectives, San Juan Hill and Kettle Hill, they began taking more fire. Spanish snipers, in their elevated position, had a clear shot at any cavalryman who stood. Casualties mounted, a half-hour passed, and still no orders arrived to attack. Finally, 1st Lt. Jules Ord of the 6th Infantry . . . yelled to his men, "Follow me, we can't stay here." Ord's charge energized the Rough Riders and parts of the 10th to join the attack. Pershing was amazed and proud at what he saw: "Each officer or soldier next in rank took charge of the line or group immediately in his front or rear and halting to fire at each good opportunity, taking reasonable advantage of cover, the entire command moved forward as coolly as though the buzzing of bullets was the humming of bees. . . .

The men waded across the San Juan River and rushed forward, slowed only momentarily by a barbed-wire fence, which most chose to climb under. In the confusion the men of the 10th divided themselves between Ord's 6th Infantry charging up San Juan Hill and Roosevelt's Rough Riders attacking Kettle Hill. Pershing found himself with the Rough

> Riders, running up the exposed slopes of Kettle Hill. It was quickly taken. In the last push to the top he saw the Spanish fleeing their positions and heading for Santiago.
>
> Pershing had a perfect view from Kettle Hill of the ongoing fight for San Juan Hill. Realizing how tenuous it was, he, and the other men on Kettle Hill, rushed forward to assist. There they struggled against the worst fire Wheeler, a Confederate Civil War veteran, had ever seen. Despite the enemy salvos, the men pushed forward, assisted by the timely arrival of a few Gatling guns brought forward for the attack. A battle yell went up along the American line. After a final, brief American artillery barrage, the troops made a final lunge for the top. Ord, with the help of the 10th Cavalry, was the first American to reach San Juan's summit, where he was immediately killed by enemy fire.[7]
>
> — U.S. Army Center of Military History

Aftermath of the Attacks on Kettle and San Juan Hills

After the fall of San Juan Hill, the Spanish quickly capitulated, setting a dangerous precedent that U.S. forces should not come to expect from a more tactically adept adversary.

> U.S. troops attacked the San Juan heights on July 1, 1898. Dismounted troopers, including the African-American Ninth and Tenth cavalries and the Rough Riders commanded by Lt. Col. Theodore Roosevelt went up against Kettle Hill while the forces led by Brigadier General Jacob Kent charged up San Juan Hill and pushed Spanish troops further inland while inflicting 1,700 casualties. . . . On July 16, the Spaniards agreed to the unconditional surrender of the 23,500 troops around the city.[8]
>
> — Library of Congress Researchers

Method of Assaulting Those Junglous Strongpoints

The first men to climb Kettle Hill did so in clusters. As they approached the foe trenches, they fanned out into separate squad-sized rows. All the while, they had been under heavy enemy fire.

> They had no glittering bayonets, they were not massed in regular array. There were a few men in advance, bunched together, and creeping up a steep, sunny hill, the top of which roared and flashed with flame. The men held their guns pressed across their breasts and stepped heavily as they climbed.[9]
>
> — *Eyewitness to History*

The lack of any particular assault method was soon apparent. Those brave souls were bent on charging headlong into whatever the enemy had to offer.

> I think the thing which impressed one the most, when our men started from cover, was that they were so few. It seemed as if someone had made an awful and terrible mistake. . . . You felt that . . . these few men were blindly following out some madman's mad order. It was not heroic then, it seemed merely terribly pathetic.[10]
>
> — *Eyewitness to History*

As during Pickett's Charge at Gettysburg, the participating Americans had only one thing in mind—to close with and destroy the enemy.

> Behind these first few, spreading out like a fan, were single lines of men, slipping and scrambling in the smooth grass, moving forward with difficulty, as though they were wading waist high through water, moving slowly, carefully, with strenuous effort. It was much more wonderful than any swinging charge could have been.[11]
>
> — *Eyewitness to History*

Yet, the magnificent courage of those advancing Americans could not alter what happened. Many were methodically picked off by the Mauser-armed defenders.

> They walked to greet death at every step, many of them, as they advanced, sinking suddenly or pitching forward and disappearing in the high grass, but the others' waded on, stubbornly, forming a thin blue line that kept creeping higher and higher up the hill. It was as inevitable as the rising

Figure 9.1: Battle of San Juan Hill
(Source: "The War in Cuba", Wikimedia Commons, public domain, image designator "Battle_of_San_Juan_Hill_-_Near_Santiago,_Cuba.jpg.")

> tide. It was a miracle of self-sacrifice, a triumph of bulldog courage, which one watched breathless with wonder.[12]
> — *Eyewitness to History*

Because the Spanish had placed their defenses along the hill's topographical crest, its contour had partially protected the attackers. A few made it to the blockhouse, and the rest is history.

> The men of the Ninth and the Rough Riders rushed to the blockhouse together, the men of the Sixth, of the Third, of the Tenth Cavalry, of the Sixth and Sixteenth Infantry, fell on their faces along the crest of the hills beyond, and opened upon the vanishing enemy.[13]
> — *Eyewitness to History*

A Legend Is Born

Those events at San Juan Heights were to create a glorified image of the U.S. fighting man that would follow him into the modern era. He had been part of a force so irrepressible as to easily advance under any circumstance. And that advance had generated so much fear that no defender could hold his ground.

> The fire of the Spanish riflemen, who still stuck bravely to their posts, doubled and trebled in fierceness, the crests of the hills crackled and burst in amazed roars, and rippled with waves of tiny flame. But the blue line crept steadily up and on, and then, near the top, the broken fragments gathered together with a sudden burst of speed, the Spaniards appeared for a moment outlined against the sky and poised for instant flight, fired a last volley and fled before the swift-moving wave that leaped and sprang up after them.[14]
>
> — *Eyewitness to History*

Just as local militia often dispersed in the face of a Redcoat charge in the U.S. War of Independence, so too had foreign regulars now run from a Yank assault in 1898 Cuba. U.S. soldiers from both sides had regularly defied death in the U.S. Civil War. And, the turn-of-the-century armchair warrior considered any fellow citizen under arms as heroic. So, the civilian-populated, top-down War Department came up with a strategy that insured not only martial victory, but also a strong economy. The indomitable spirit of the American fighting man would be combined with new equipment and charismatic leadership to carry every military engagement. Just as Teddy Roosevelt had thrown caution to the winds and his troops followed, so could all future battles be won the same way. Yet, with the average American now believing Teddy and his 1st Volunteer Cavalry had gone up San Juan Hill, the official version of the Spanish American War became suspect.

A Different Interpretation of What Occurred

There's another way of looking at those events near Santiago on 1 July 1898. U.S. forces had advanced toward a defensive line within full view of opposition soldiers who all carried a rapid-firing rifle that could accurately shoot up to 2000 meters with a smokeless round.

> The rifle's sturdy, ladder-style rear sight was graduated to 2,000 meters, the front sight involving a blade that was drift-adjustable for windage.[15]
>
> — *The American Rifleman,* February 2002

Not only had Teddy Roosevelt stopped just short of San Juan Hill, but the Spanish defenders of Santiago had been outnumbered 16 to 1.[16]

> On July 1, 1898 . . . a battle occurred that would be one of the turning points in . . . U.S. military history. . . . During the American assault on the strategic Cuban city of Santiago, 750 Spanish troops defended positions on San Juan and Kettle hills. The attacking force numbered approximately 6,600 American soldiers, most of them regulars, armed with the then-new smokeless-powder Krag-Jorgensen rifle and supported by artillery and Gatling gun fire.[17]
>
> — *Shooting Times,* 2017

> When the brave—if disorganized—assault was over, the Americans discovered . . . they had suffered more than 1,400 casualties. That was almost 20 percent of their attacking force. To have such losses inflicted upon them by such inferior numbers was a shock. . . . [T]he investigation laid the blame . . . on the . . . Spaniards' Modelo 1893 Mauser rifles.[18]
>
> — *Shooting Times,* 2017

After being badly hurt by Spanish cannons and long-range Mausers, Roosevelt's 1st Volunteer Cavalry and the 10th Regular Cavalry (containing the future Gen. Pershing) had charged the last 150 yards up Kettle Hill within full view of the enemy. Only saving them from disaster had been Spanish positions along the topographical crest of the hill instead of its military crest. (From the latter, one has an unobstructed view beyond grenade range). While the defenders would eventually forfeit every part of San Juan Heights, they only took half as many casualties as the Americans.[19]

So, while the Cuban victory had been speedy and decisive for the U.S. Army, it had come at great cost. Well into the modern-age, American doctrine would require 3:1 odds or better for any infantry assault.

There were other problems with the Yank effort. The enemy 's standard issue rifle could not only fire more quickly than the U.S. counterpart, but also had a maximum range of 4000 meters.[20] When the U.S. troops made little attempt to disguise their approach, all Spaniards manning the defensive lines could have taken them under volley fire up to two and a half miles away. And a big part of those 1400 U.S. casualties could have been avoided through more ground element maneuver.

Questionable Parts of U.S. Assault Method

The Americans had landed at Daiquiri and Siboney and then marched eastward through Las Guasimas toward Santiago. While such a direct approach to the final target had gained a rapid victory, American forces suffered many losses en route. Several issues point to strategy problems.

Too little emphasis on individual and small-unit maneuver was evident from mixing quickly trained volunteers with regulars. Apparently assumed was that enough brave young men crossing the last 100 yards would get the job done. (See Figure 9.2.)

But, not enough thought had gone into the unique situation those troops would face—semi-jungly terrain manned by people with rifles and cannon that could shoot a long way. When first engaged, U.S. forces deployed on line in vegetation too thick to permit such a formation. They marked their approach route with an observation balloon. They crossed a river at a ford that had been preregistered for enemy artillery. And then, they ran an impromptu assault only saved by its pell-mell nature. While the enemy's volley rifle fire had been the equivalent of a Gatling gun, U.S. leaders saw no real need to avoid it.

The Americans also experienced land navigation and scouting problems. Quartermaster Pershing had been assigned the job for the 10th Cavalry. That distinguished outfit had probably used Indians for this purpose on the Western frontier after the Civil War. And, it would partially revert to the Indian way of fighting during

Figure 9.2: America's Fighters Were Determined but Mortal
(Source: "The War in Cuba", Wikimedia Commons, public domain, image designator "Santiago_-_Capture_of_San_Juan_Hill (1).jpg.")

its assault up Kettle Hill. Small bunches of troops had begun the climb separately. Then, each bunch fanned out near the top for more firepower.

What Would Henceforth Be Required of U.S. Infantrymen

Like the British regulars of 1780, American ground-pounders at the turn of the 20th Century were still expected to remain upright during the final stages of any assault. They were forced to ignore how little cover they had from enemy grenades, mines, automatic-weapons fire, and artillery impacts. Stopping to help an injured comrade was also forbidden. And, even crawling forward took special permission.

With the invention of the machinegun, this Western infantry tradition was about to play havoc with the Allied casualty counts of WWI.

Few Changes in the Strongpoint Thread

During the Spanish-American War, what had been planned as a broad linear attack on the primary enemy strongpoint atop San Juan Hill soon turned into a number of tiny, separately operating assault lines. While their randomly generated piecemeal effort then created a certain amount of enemy surprise, it was never incorporated into the U.S. assault method. Something like it did reappear in the Asian Communist attacks at the Chosin Reservoir in Korea and Con Thien in Vietnam.[21] But in both cases, separate Stormtrooper squad columns would take the place of the tiny rows of regularly trained riflemen.

10 World War I

- Would the Yanks pick up any bad habits from their Allies?
- Which new maneuvers did the Americans attempt?

More than enough Doughboys were now available to fight.

(Source: "American Troops Advancing," by Harold Brett, U.S. Army Center of Military History, artphoto archives, illustration designator "avop05-98_1.jpg,")

How the Huns Were to Operate

Early in the "Great War," the Germans protected their riflemen with trenches and their machinegunners with deep pits. Soon, they realized too many riflemen were succumbing to bombardment,[1] and too many machinegunners to capture. During a pre-assault barrage, both groups more productively shifted into nearby shell holes. From there, they could unexpectedly shoot into the advancing waves of Allied infantrymen.

Each shell hole held roughly a squad, so the Hun rifleman's role became safeguarding a crew-served weapon.[2] From hastily fortified pits,[3] such weapons could more effectively operate. Yet, those now reinforced German squads had been encouraged to move slightly forward, sideways, or backwards depending on the situation.[4] Their loose array of positions was to bend without breaking, and thus became the first elastic defense in depth. By April 1917, what had started out as a random series of occupied shell holes became a carefully arranged matrix of bunkers. (See Figure 10.1).[5] The Huns had thereby made the logical transition from a linear defense in depth to a belt of mutually supporting strongpoints.[6]

Before long, it became evident that reverse-slope strongpoints were nearly impossible for Allied gunners to target.[7] If camouflaged, they also eluded aerial observation.[8] And when properly staggered, each could be protected by the crisscrossed fire of two to its rear. That lead curtain not only made the bastion harder to assault, but also possible to reinforce, resupply, or abandon. Even behind the main line of resistance (MLR), each little fortress could maintain its own 360° defense. At first, its occupants were to hold out until relieved by counterattack. Then, it became apparent that a matrix of tiny bastions is just as durable when a few shift rearward. A gradually receding defensive belt needs no coordinator, as long as the noncommissioned officer {NCO) in charge of each fort decides when to withdraw.[9]

Near the end of WWI, the Germans also developed a 3GW-style offense. Luckily for the Allies, it gained ground too quickly to be properly supplied.

> The watershed of modern military tactics is WWI. . . . The other side . . . adopted specialized formations for attack by infiltration (so-called Hutier tactics) and follow-up by regular formations. Its tactical success . . . led to *"blitzkrieg."*[10]
>
> —U.S. Defense Advanced Research Project Agency

The goal of such an attack was to bypass most resistance and move forward as far as possible.[11] British machineguns hardly fired a shot at the lead elements of the Germans' Spring Offensive in 1918.[12]

> The first wave was an infantry probe (from the accompanying division) whose purpose was to identify enemy positions for the next wave, about 250 meters behind. The second wave consisted of the elite storm companies and the flame-

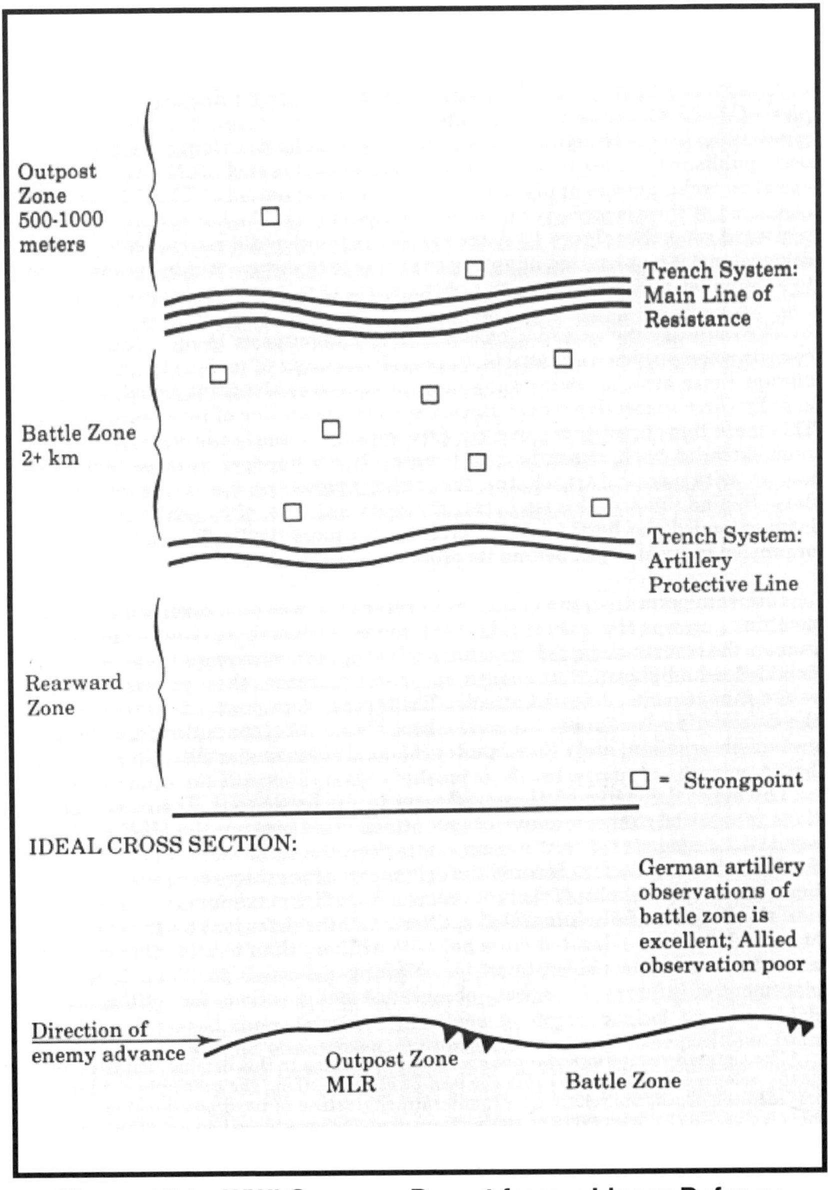

Figure 10.1: WWI Germans Depart from a Linear Defense
(Source: *The Changes in German Tactical Doctrine during the First World War*, U.S. Army Combat Studies Inst., Leavenworth Paper No. 4 (1981), p. 14)

> thrower section, with additional infantry support from the division. This second wave attempted to penetrate the enemy zones by pushing through weak areas to envelop enemy positions. Supporting these efforts was the third wave, about 150 meters behind, which contained the storm battalion's heavy weapons and similar additional support from the division.[13]
>
> — U.S. Army, *Leavenworth Papers No. 4*

At the front of those companies were squads with recently rehearsed Stormtrooper technique. (See Figure 10.2.) Through "reconnaissance pull," each of those squads then conducted a "quick deliberate attack." But, even highly skilled forward elements will eventually run out of ammunition.[14]

To penetrate Allied lines with an indirect-fire deception, each German squad had only to: (1) crawl up to the enemy's barbed wire; (2) insert a bangalore torpedo; (3) detonate that torpedo during a precision artillery barrage; and then (4) assault with concussion

Figure 10.2: Assault Load of Early Stormtrooper

(Source: "Bild eines Sturm-Pioniers vom Sturm-Bataillon 5 Rohr," by L.Dettmann, https://commons.wikimedia.org/wiki/File:StB5_%E2%80%93_Sturm-Pionier.jpg#filehistory.)

grenades and bayonets after the barrage was shifted. To keep most defenders from realizing a ground attack was underway, the Huns fired no small-arms.[15] Then, to outflank an Allied machinegun, some Stormtroopers moved sideways a short distance.

> The essential elements of the tactics that [Capt.] Rohr developed in the course of these experiments were (1) the replacement of the advance in skirmish lines with the surprise assault of squad-sized "stormtroops," (2) the use of supporting arms . . . coordinated at the lowest possible level . . . to suppress the enemy during the attack, and (3) the clearing of trenches by rolling them up with troops armed with hand grenades.[16]
>
> — *Stormtroop Tactics,* by Gudmundsson

But, not all tactical successes are strategically helpful. With too many Doughboys now in the way, the Huns experienced supply shortages and raging influenza. And the German's tactical superiority was squandered on areas that were not Allied centers of gravity (like command centers or supply dumps).[17] Then, there was no stopping the waves of fully exposed Doughboys.

> By the final 100 days [of the war], German forces just could not withstand the overwhelming artillery, firepower, and mass of the rejuvenated Allied forces.[18]
>
> —U.S. Defense Advanced Research Project Agency

How the British Were to Operate

The British entered WWI expecting to outmaneuver the foe across the coastal plains of northwestern Europe. They soon discovered no safe place above ground. (See Figure 10.3.) Rows of enemy trenches had been too hard to breach.

> [I]n 1914. Europeans sleepwalked into war dreaming of cavalry charges and massed infantry charges with fixed bayonets. They awoke to confront the machine gun, . . . the tank, and the airplane.[19]
>
> — *National Interest,* n.d.

Most soldiers in the [British] Army were in the infantry. They were foot soldiers, trained to march and carry all their equipment. A soldier had to be dressed at all times with his rifle ready.

Infantry soldiers lived in trenches for up to weeks at a time. When an attack was ordered, lines of infantry soldiers climbed out of the trenches and marched towards the enemy. This was called “going over the top”. . . .

Sappers planned where trenches should be dug, miners dug tunnels. . . .

Figure 10.3: Just Looking Out of a Trench Was Risky
(Source: Wikimedia Commons, public domain, image designator "WW1_sniper_sketch_(35438988912).png."

> Despite the many volunteers who joined the armed forces, there was still a shortage by 1916. A law was passed forcing men to join up, called conscription. By 1918 all [English] men under 51 were being called up.[20]
>
> — *British Broadcasting Corporation,* n.d.

By 1918, the British were attacking differently than before. More precise artillery was aimed at better hidden targets. But, it would be Allied armor to permit most ground progress.

> [As] at Neuve Chapelle . . . the first attempt at breakthrough in March 1915. . . [t]his hurricane style barrage was effective as 288 pounds of artillery high explosive per yard of trench pummeled the Germans. However, the British [later] made the mistake of expanding the concept of hurricane barrages to encompass a wider area for a longer period of time—a linear extrapolation that ignored the declining pounds of artillery per square yard of trench.[21]
>
> —U.S. Defense Advanced Research Project Agency

> [The German Maxim MG 08 machinegun] could reliably spray out 500 rounds a minute, which cut down massed Allied infantry assaults like a scythe. . . .
>
> The machine gun's nemesis was the tank. Entrenched machine guns could survive weeks of massive artillery bombardment at muddy hells like Passchendaele. But a tracked vehicle, well-armored enough to survive machine gun fire and powerful enough to knock down barbed wire, could breach a path for attacking infantry to cross No Man's Land without being cut to pieces.
>
> These first tanks were hot, noisy and unreliable. . . . They were also vulnerable to German artillery firing directly over open sights. Nonetheless, they were powerful enough to help end the bloody trench stalemate of World War I.[22]
>
> — *National Interest,* n.d.

Nor on defense could the British ever equal the Germans. Their still linear formations lacked the firesacks that had been created by separate Hun strongpoints.

The British also imitated the defense in depth concept, but

relied on static independent fortifications that lacked the flexibility and mutually supportive fields of fire of the German defenses.[23]

—U.S. Defense Advanced Research Project Agency

How the French Would Fight

Between 1914 and 1918, French army casualties in WWI would total 1.4 million killed, 4.3 million wounded, and 537,000 taken prisoner or missing, according to the Public Broadcasting System (PBS). Of all soldiers who had taken part in the fighting, this casualty rate approached 73.3%. A few French commanders eventually shifted away from the massive attacks. Marshall Henri Petain developed more flexible and mobile forms of warfare that were to somewhat influence the Americans.[24]

Most of WWI had occurred within trench systems for which boundaries changed very little. French generals like Joseph Joffre had tried to break the stalemate with massed infantry. His troops would walk toward enemy positions and then break into a collective charge at the last moment, according to the French Infantry Tactics Website. A typical assault might take three and a half minutes—ample time for the German machine gunners to kill every participant.[25]

To further support the infantry, the "creeping barrage" came into practice during the French victory at Verdun. This early struggle had lasted from February to December 1916 and claimed 700,000 lives, says historian Khushaal Vyas. Artillery gunners fired on enemy positions, and small infantry units followed right behind the exploding shells. As the ground pounders reached a trench, the artillery ceased, and bayonets or grenades were used against its defenders. So doing must have sounded to other Germans like a continuation of the bombardment. British reinforcements then helped to secure the area.[26]

Western Front commander Petain preferred speed and firepower over massed attacks. He subdivided infantry units into smaller and more mobile teams that could keep enemy defenders pinned down with machinegun fire, according to the French Tactics Website. Then, other teams threw grenades during a bounding maneuver to overcome any resistance. But, a few subordinate commanders had stuck with the "walking-fire approach" where an entire unit continues to shoot at enemy positions—a method that the British and Australians were to mirror.[27] And in all likelihood, the Americans would copy it as well.

The Americans' Training by Their Allies

Upon reaching Europe, the Doughboys were mostly instructed on trench warfare by the English and French. But, they spent far more time on building and defending trenches, than how to attack one.

> Specialized schools and camps for [American] artillery, infantry, and the other arms and services were established. French and British veterans provided training in these camps. Men learned to use gas masks, dig trenches and lay barbed wire. . . .
>
> [T]he average soldier had six months of training in the United States, two months of training in France, and one month in a quiet sector of the front.[28]
>
> — America Battle Monuments Commission

After the individual instruction came a little combined-arms unit training. Likely consisting of an on-line tank/infantry assault behind a rolling artillery barrage, it would do better in theory than in practice. But the Americans complied, because they knew the trenches had to be trans-navigated.

> For those who survived their first month in the trenches, the battalions returned to their divisions to train as an all arms team. This phase reflected the [American Expeditionary Force's] AEF's focus on "open warfare." Gen. John J. Pershing and his staff felt the war would only be won when the enemy was forced out of the stalemate of the trenches and defeated in a battle of maneuvers in the open. The AEF grew from the first four divisions to a total of 42 sent to France. Between the ports of Calais and Boulogne on the English Channel, Americans trained in 13 different British training centers, and more than 50 French training areas across France.[29]
>
> — America Battle Monuments Commission, n.d.

The Americans' Battlefield Experience

During WWI, some 53,000 Americans would be killed within a relatively short period of time. The first fell in the summer of 1917, but most wouldn't get deeply enough involved until 1918. Then, they either died countering the Germans' Spring Offensive or in mid-July during the Second Battle of the Marne. Nearly 26,000 would make the ultimate sacrifice during the final 47-day Meuse-Argonne Offensive of 26 September 1918.[30]

The Yanks knew why the trenches had to be bypassed, but not why they had been dug. They had never before faced perfectly integrated machinegun and artillery fire. Even during the 100 Days breakthrough of 1918, they were unable to evade its devastating effects.

Then, something illogical happened within the U.S. military. Only blatantly obvious automatic weapons and artillery would be targeted before and during any major attack. The rest would be ignored, as the unavoidable price of invasion. And throughout this and the next world war, large U.S. formations would cross contested battlefields on-line, fully erect, and in broad daylight. No one knows whether a mental block from WWI had been responsible for the overly risky behavior or some other factor. But, just as in the U.S. Civil War, American troops would continue to endure incoming fire. Had a superiority complex from the "War to End All Wars" blunted their natural degree of caution?

> One finds arrogance and the conviction of superiority vis-à-vis America's allies and [her] enemies [in WWI], along with the assumption that they had nothing to teach U.S. forces. . . .
>
> The implication was that the [French and British] . . . were lacking in character and were almost unmanly in their taste for trenches and the various accoutrements associated with trench warfare, chief among them machineguns and massed artillery, but also grenades and mortars.[31]
>
> — *War on the Rocks,* August 2018

Upon arriving in France, WWI Marines had considered themselves so imposing as never having to dig in. They soon learned a tragic lesson. The two sides had been locked in a murderous stalemate because of the available weaponry.

Massed machinegun and artillery fire made virtually suicidal the open-field maneuvers that British, French, and German commanders had dreamed of in 1914, and Pershing still pursued in 1918. Such a push couldn't advance beyond the range of its own artillery.[32]

Turning the Tide

Heavily depending on the near-total surprise of a Stormtrooper assault, the Germans launch their Spring Offensive in May of 1918.

> The Germans reach[ed] the Marne at Chateau-Thierry, and the road leading to Paris seemed open. On 1 June, 2nd [U.S.] Division formed a line across the road, the Marine brigade in the center, and the Army's brigade regiments . . . on the flanks.[33]
>
> —*The Marines*,
> from Marine Corps Heritage Foundation

But, the U.S. Marines are unimpressed, and the legendary marksmanship of their low-ranking riflemen threatens to dampen enemy resolve at 800 yards.[34]

> The Germans attacked [for four days]. . . . Then on 6 June, the 2nd Division counterattacked. In front of the Marines stretched . . . Belleau Wood. In the first attacks, Marines in well-dressed lines crossed open wheat fields. . . . Once into the woods, the fragments of Marine lines instinctively formed into small combat groups.[35]
>
> — *The Marines*,
> from Marine Corps Heritage Foundation.

Belleau Wood

Like at Gettysburg, the Marines are forced to cross half a mile of open wheat field to reach Belleau Wood. This brings into question whether 5199 U.S. casualties were really necessary in that battle. (See Map 10.2 and Figure 10.4.)

Gen. [Charles] Krulak [while Marine Commandant] said he took his first walk a year ago, starting near the town of Lucy-le-Bocage, where the WWI marines launched their attack June 6.

"I walked toward the tree line through waist-high wheat, just as they did 80 years ago," the Commandant said. "His-

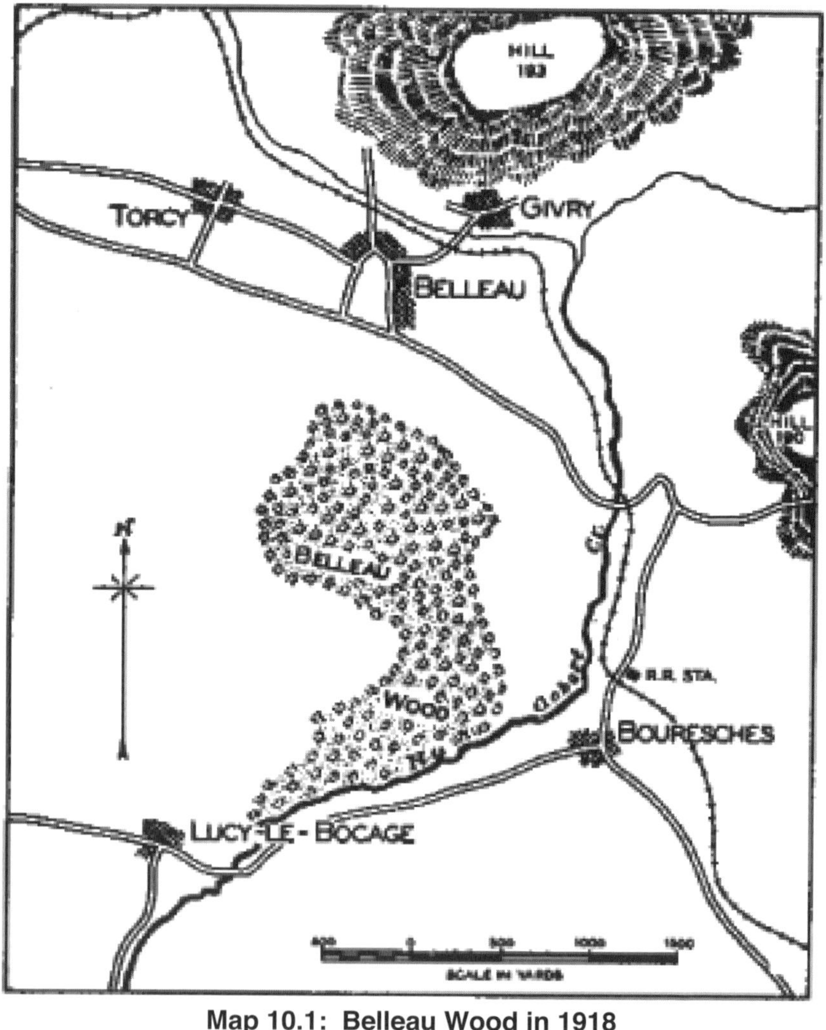

Map 10.1: Belleau Wood in 1918
(Source: FMFRP 12-2, "Infantry in Battle," p. 31.)

tory books describe that 800-yard advance, but I never fully appreciated it until I walked it myself. The Germans had every square inch of that field covered by machinegun and artillery fire. The Marines paid dearly with every step they took."[36]

– *Leatherneck Magazine*, August 1998

This [Thursday, 6 June] would be the most catastrophic day in Marine Corps history. More Marines would be killed and wounded on this single day than in all its previous existence. . . . For all intents and purposes, the old warriors of the U.S. Marine Corps were virtually wiped out, mainly leaving replacements to carry the war to its final term. . . .

At 1700 [hours] five-hundred Marines of the 3rd Battalion, Fifth Marines, jumped off . . . with an almost perfect skirmish line. . . .

What was not known officially, mainly because no American reconnaissance had been undertaken beforehand, was that the woods were chock full *[sic]* of Germans and the approaches were thoroughly covered by their artillery. In addition, and more devastating to the Marines, were the many heavy Maxim [machine] guns that had been placed in advantageous, and interlocking [fire] positions within the wood. . . .

As soon as 3/5 left its wooded place of safety to cross the open wheat fields lying before them, deadly machine gun fire was encountered. . . . While crossing the open ground to their immediate front the losses to Berry's battalion [3/5] were terrific. . . .

In just a few short minutes the Third Battalion, Fifth Marines, was just a memory. . . .

According to its semi-official history, 3/6 [3rd Battalion, 6th Marines] jumped-off at exactly 1700 from their positions . . . just to the east of Lucy-le-Bocage. . . . In fact, in very short order they ran into a wall of Maxims that 3/5 had [also run into]. . . .

Both the 84th and 97th [companies of 3/6] advanced through open wheat fields [south of Belleau Wood] under constant observation and direct artillery and machine gun fire. . . .

Figure 10.4: Unhappy Wheat Field Survivors Enter the Wood
(Source: Public-domain painti4g by Georges Scott [1873-1943] illustration for French magazine from https://commons.wikimedia.org/wiki/File:Scott_Belleau_Wood.jpg.)

Holcomb's battalion [2/6] received orders at 1630 to jump-off at 1730 with one company. . . .

At 1730 the 96th went forward as ordered. . . . Going through the wheat field [south of Belleau Wood toward the town of Bouresches], against emplaced, heavy machine guns, and the dreaded shell fire, the losses of the 96th were horrendous. . . .

The 6th of June 1918, was . . . "The longest Day," . . . for the 4th Marine Brigade. The losses for the day, thirty-one officers and 1,056 enlisted men. . . . The largest proportion were suffered in the afternoon among the three battalions engaged, 3/5, 3/6, and 2/6.[37]

– *Their Time in Hell*, by Clark

Just getting into Belleau Wood had cost the U.S. Marines 1000 dead and wounded, more than from all previous engagements in their history. But, its unexpected price took a while to sink in.[38]

> On the morning of 18 July, the 2nd Division loaned to the French went in south of Soissons, and by nightfall the 5th Marines had taken the village of Vierzy. The following morning, the 6th Marines continued the attack, crossing an open wheat field behind a thin line of French tanks. The two days of fighting cost the Marines two thousand dead and wounded.[39]
>
> —*The Marines,*
> from Marine Corps Heritage Foundation

While the Germans would never attack again, the "nineteenth century [Allied] tactics and raw courage" had left a lot to be desired.[40]

What Then Led to the End of the War

While Allied tanks couldn't do much to protect accompanying infantrymen on a open field, they made all the difference in a trench network. (See Figure 10.5.)

> The German Spring Offensive came close to breaking the Allied front line. . . . In the Second Battle of the Marne (15 July-6 August), the Germans . . . failed to deliver a decisive blow and on 18 July the Allied counterattack [is launched], led by the French.[41]
>
> — Imperial War Museums of Britain

Later versions of the British Mark I tank could cross barbed wire and trenches without bogging down. That's how the Allies finally broke through the Huns' fortified network.

> The Germans evolved an extremely elaborate defense system using pillboxes, i.e., concrete shelters for machine guns. Behind the pillboxes were more lines of barbed wire and more trenches and dugouts reinforced with concrete to withstand artillery bombardment. Behind these defenses

Figure 10.5: The British Mart I Tank
(Source: Wikimedia Commons, public domain, image designator "A_Mark_I_tank_RMG_PU9983.jpg."

> were still more lines of trenches that were effectively out of range of the enemy's artillery fire. By 1918, the Germans had constructed some trench systems that had a depth of 14 miles (22 km). . . .
>
> The Allies' increased use of the tank in 1918 marked the beginning of the end of trench warfare, however, since the tank was invulnerable to the machine gun and rifle fire that were the trenches' ultimate defense.[42]
>
> — *Encyclopedia Britannica,* n.d.

The Allies' counterattack gained momentum in August of 1918 when their newest armor appeared on the scene in great numbers.

> At 4:20 AM on 8 August 1918, the Battle of Amiens began. It was a morning of heavy fog and the Germans were taken completely by surprise. . . . The Australian Corps and Ca-

> nadian Corps spearheaded the attack and advanced quickly behind the 534 tanks.[43]
> — Imperial War Museums of Britain

The Germans adjusted their fire plan, the tempo of Allied operations slowed to a grinding motion. But, war materiel shortages had begun to bother the Germans.

> When the advance was halted on 11 August, the Allies shifted their attack to a different part of the line. . . . The Allies continued to attack in this way throughout the summer and autumn of 1918, giving the . . . exhausted. . . German Army little respite.[44]
> — Imperial War Museums of Britain

> By the end of August the Allies had notably captured Albert, Bapaume, Noyon and Peronne during the Second Battle of the Somme.[45]
> — Imperial War Museums of Britain

While now advancing, the British and French had lost huge numbers people over the previous four years of war. And, there was now an abundance of Americans to throw into the breach.

> By [then] . . . there were over 1.4 million American troops in France. It was the arrival of these fresh troops that enabled the Allies to continue fighting after their significant losses during the German Spring Offensive.[46]
> — Imperial War Museums of Britain

> The attack on the St. Mihiel salient (12-15 September) was the first and only American led attack during the First World War. It was a relatively easy victory as it caught the German Army on the retreat but it established the American Army as a formidable fighting force.[47]
> — Imperial War Museums of Britain

Battle of St. Mihiel Salient

The U.S. units wouldn't maneuver much at the Battle of St.

Miheal. With great resolve, they just kept coming as the Germans gradually withdrew. But, the U.S. casualty count speaks volumes with regard to how much danger they encountered.

> The Allied attack against the Saint-Mihiel salient provided the Americans with an opportunity to use their forces on the Western Front *en masse.* Although lacking some of the tactical skills of the French and British, the U.S. First Army carried the day through sheer determination and its multifaceted plan of attack. The battle was also noteworthy as the first major use in the war of the U.S. Army Air Service . . . led by William "Billy" Mitchell and the aggressive tank assaults by George Patton. . . .
>
> The attack on the Saint-Mihiel salient on 12 September gave him [Pershing] the opportunity to use the U.S. First Army in combat for the first time. The American part of the assault was to be conducted by two "super" corps, each with three divisions in attack and one in reserve. Two smaller French corps would provide support on the western part of the salient.
>
> General Erich von Ludendorff—now short of men and aware of the coming Allied offensive—had decided to withdraw from the salient to a shorter and more easily defended line to the rear. As the Germans were withdrawing, the Allies attacked. With much of their artillery not in place, the Germans were poorly prepared to maintain the front line, an advantage that the attacking Americans were quick to exploit. The relative ease of the initial American attack came as a surprise to Pershing, and he sent orders to his commanders to speed up their advance. By 13 September lead units of the U.S. First Army had met up with Allied troops advancing from the west. Three days later, the offensive was halted, with the salient in Allied hands.
>
> Losses: U.S., 7,000 casualties; German, at least 17,500, including 10,000 captured.[48]
>
> — *Encyclopedia Britannica,* n.d.

What Was Changing in the Way WWI Was Fought

Besides having improved tanks, the Allies now controlled the sky above the battlefield. Their aerial spotters could better assess German intentions and then hit with artillery any point target. But, many of those targets were now well camouflaged. So, hidden German automatic weapons could still make any daytime ground advance very costly. And, all Western commanders accepted their presence as inescapable.

> The Allied armies deployed new tactics to overcome the stalemate of trench warfare. Artillery, tanks and air power were successfully utilized in a new coordinated all-arms approach. Allied success saw fighting move from the trenches out into the open.[49]
>
> — Imperial War Museums of Britain

> Allied artillery dominated the battlefield paving the way for a breakthrough. However, German machine guns hindered their advances so that most attacks were . . . [initially started] under cover of darkness.[50]
>
> — Imperial War Museums of Britain

> Tanks were still relatively new weapons and were most useful for crushing barbed wire obstacles, destroying machine-gun posts, and in village fighting. They would be followed by small groups of infantry. They carried cribs, frames made of wood and steel, which could be dropped to enable them to cross wide trenches.[51]
>
> — Imperial War Museums of Britain

Then Came the Deadly Meuse-Argonne Offensive

Somewhere within the final engagement of the "Great War", American forces were about to lose another 26,000 men. (See Map 10.2.)

> With the success at St. Mihiel, the Americans were moved to support the ambitious attack planned by Marshal Foch at the Battles of Meuse-Argonne. This was the main contribution of the American Army in the First World War and the losses were high amongst their inexperienced troops.[52]
>
> — Imperial War Museums of Britain

> Pershing's . . . (AEF) was to play the main attacking role, in what would be the largest American-run offensive of World War I.[53]
>
> — *History Channel,* n.d.

After 400,000 U.S. troops were transferred from St. Mihiel, the Meuse-Argonne offensive got underway.[54]

> At 5:30 on the morning of September 26, 1918, after a six-hour-long bombardment over the previous night, more than 700 Allied tanks, followed closely by infantry troops, advance against German positions in the Argonne Forest and along the Meuse River.[55]
>
> — *History Channel,* n.d.

> By the morning of the following day, the Allies had captured more than 23,000 German prisoners; by nightfall, they had taken 10,000 more and advanced up to six miles in some areas.[56]
>
> — *History Channel,* n.d.

Reaching the Hindenburg Line

The Huns' original defense zone comes next. The whole system had been laid out linearly, as opposed to the mutually supporting bastions of a network. Australian, British, French and American forces participate in an attack that begins with a 56-hour long bombardment, using 1,637 guns. In the last 24 hours of the siege, 945,052 British artillery shells would land, to include a creeping barrage at St. Quentin Canal.[57]

> By late September the Allied forces were facing the Hindenburg line, a series of heavily fortified positions that formed the main German defenses.[58]
>
> — Imperial War Museums of Britain

> The Battle of St. Quentin Canal (29 September 1918) was a crucial victory that broke through one of the strongest sections of the Hindenburg Line. Following the complete

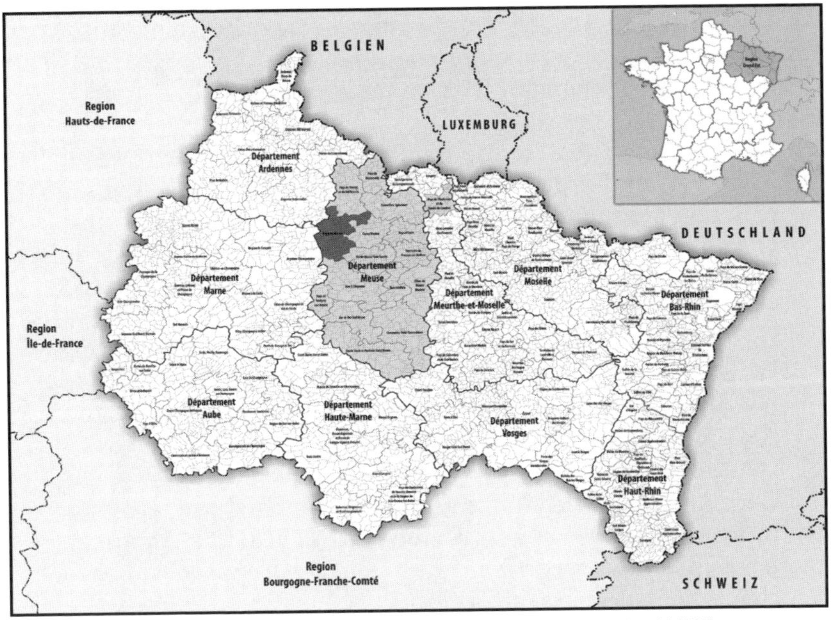

Map 10.2: Argonne Forest Took Its Toll in WWI
(Source: Wikimedia Commons, public domain, image designator "Gemeindeverband_Argonne-Meuse_2018.png.")

> breakthrough of the line in early October, General Ludendorff [began to think about armistice].[59]
>
> — Imperial War Museums of Britain

The Allied Counterattack Moves On

After breaching what the Germans call the Siegfried Line, the Allies return to the push at Meuse-Argonne.

> [I]t [the battle of Meuse-Argonne] was renewed again . . . on October 4. . . . [T]he German troops held on another month, before beginning their final retreat. Arriving U.S. reinforcements had time to advance some 32 kilometers before the general armistice was announced on November 11.[60]
>
> — *History Channel*, n.d.

It would be in wooded terrain over the next terrible month that American forces were to suffer the large clump of casualties. The German's new strongpoint belt was likely responsible.

> Following the German retreat from the Marne River in July, Gen. Ferdinand Foch . . . [had] designed a series of convergent and practically simultaneous offensives against the shaken German armies. One was a joint operation in the Meuse valley toward the Mézière and Sedan rail centre. The Americans . . . faced the most difficult natural obstacle, the dense Argonne Forest. Gen. John Pershing's opening surprise attack advanced 5 miles . . . along the Meuse River but only 2 miles . . . in the difficult Argonne Forest sector. Attack after attack edged deeper into the Germans' defensive position. On the 11th day of the American offensive, the Germans recognized that they were outflanked and retreated to avoid capture. . . . By October 31 [1918] the American forces had advanced 10 miles . . . and the Argonne had been cleared of German troops.
>
> Hard fighting continued in the Meuse-Argonne sector during October. More than a million Americans participated in the battles, but the American Expeditionary Force's casualties were heavy, and its largely inexperienced formations were becoming increasingly disorganized. On November 10 the Allies reached Sedan and cut the rail line there.[61]
>
> — *Encyclopedia Britannica,* n.d.

America's linear formations may have become somewhat less organized, after subdividing a little to handle enemy 3GW initiatives. But, that less overall control was productive over the long term never occurred to a headquarters that had never wanted to practice 3GW itself.

And something else may have led to the high number of friendly casualties. Head-strong U.S. formations may have stayed barely abreast of the fire pattern from a "rearward-moving" enemy defense zone. In the process, they had subjected their troops to its killing power. After all, organized withdrawal had been an integral part of the new Stormtrooper defense,[62] and "retreat combat" would be practiced by Axis powers throughout WWII.[63]

> The rolling hills and densely wooded forests of the re-

gion provided the Germans with ideal defensive positions on which to unleash devastating machine gun and artillery fire upon the advancing American forces.

The Americans faced a formidable adversary whose guns would not be silenced with ease. Nevertheless, the determined American offensive forced the Germans to continuously withdraw northward, before the armistice abruptly ended the campaign and the war.[64]

— American Battlefields Commission, n.d.

The Flaw in the Allies Final Attack Formula

As WWI progressed, Allied artillery supposedly became well enough adjusted to pave the way for infantry. Some historians even credit shelling for ultimate victory. But, in the Allies counterattack of 1918, too many rolling artillery barrages were attempted at once. In only a few places were they sufficiently concentrated to reach every machinegun dugout. Nor could a rolling barrage be slowed down enough for its infantrymen to outmaneuver an active gunpit. To keep up with the advancing barrage, their parent-unit line was forced to disregard incoming automatic weapons' fire.

Those rolling barrages had only seemed to work because of the rearward movement of the Huns. But tiny forts had been abandoned, and others still covered their approaches by fire. That's when lot of jubilant Doughboys got shot by surprise.

A rearward-moving strongpoint matrix can wreak almost as much havoc as one that is stationary. It's helping to deplete the foe's combat power. The innovative Huns had used such a maneuver to preserve limited assets.

> [M]ore than 600,000 American and French troops supported by almost 5,000 artillery pieces, 500 tanks, and 500 airplanes slowly forced their way [probably on line] through strong German defenses [at Meuse-Argonne].[65]
>
> — Library of Congress, n.d.

A review of the "Hell Fighters from Harlem" painting claims the U.S. formation "flowed around" the German machineguns, like a native American attack would have. At that point, adjacent elements supposedly destroyed those guns from the side.

Figure 10.6: Attack of the "Hell Fighters from Harlem"
(Source: Wikimedia Commons, public domain, image designator "Meuse-Argonne,_26_September–1_October_1918.jpg."

> The 369th Infantry fought valiantly in the Meuse-Argonne as part of the French 161st Division. Attacking behind a fiery barrage, the 369th Infantry assaulted successive German trench lines and captured the town of Ripont. On 29 September, the regiment stormed powerful enemy positions and took the town of Sechault. Despite heavy casualties, the 369th, called "Hell Fighters" by the French and Germans, relentlessly continued the attack at dawn. Raked by enemy machine guns, they assaulted in the woods northeast of Sechault, flanking and overwhelming enemy machine gun positions.[66]
>
> — Caption to famous WWI painting

But, the painting itself tells a different story. (See Figure 10.6.) It shows a ragged line of Hell Fighters being forced into a now defunct gun. In other words, the integrity of their formation had taken precedence over the safety of individual members. Its whole purpose had been enough personnel at the end of the assault to consolidate all ground gains. And from what the regimental commander could see, the massed firepower of such a formation met all expectations. But any U.S. Private who found himself in front of a still lethal gun was not nearly as happy. And U.S. headquarters would continue to ignore his perspective well into the future. In Vietnam, American infantry sweeps seldom took the time to investigate any ground irregularity.[67]

The Deadly Illusion

During those last difficult days of 1918, U.S. forces proved instrumental in winning the "Great War." But, they had done so without gaining fire superiority over the crew-served weapons of their opponent. Those guns would end up killing 553 Doughboys a day and injuring another 2128 during the same 24-hour period. But prideful veterans and their equally august organizations would later believe every way to avoid that much bloodshed had been explored.

In effect, a military mystery had occurred that would baffle Western experts well into the 21st Century. Those WWI Americans had been doing what their more experienced Allies recommended. So, that advice must have been somehow remiss. Britain's inherent caste system (with the "gentleman's gentleman") couldn't have added much to that formula. Nor would the French army's stratification after Napoleon's loss at Waterloo. That same army was to mutiny in 1917. And, by that time, one out of every three males in Scotland had already been lost by the English army.[68]

After defeating the British in the Revolutionary War, and Spanish in 1898, the Americans had entered WWI with considerable disdain for the European way of doing things. But, in their eagerness to help against Hun aggression, they had gone along with a British and French myth—that fully exposed human beings can directly assault a good enemy defense without taking excess casualties. The Asians had known this for centuries, and the group-oriented Germans had learned it from the Japanese. But, this common-sense axiom would be hard for Americans to embrace. Their burgeoning military bureaucracy had begun to resist all changes, however useful.

Another Shortfall That Survived the War

U.S. defense lines during WWI were identical to those of today—a long string of fighting holes, preceded by rows of barbed wire and bands of interlocking machinegun fire. The German Stormtrooper squads had easily penetrated such an array by complete surprise. But, first to die had been any American realizing the ruse. So, headquarters never acknowledged it, nor the post-war manuals. And U.S. troops would continue to defend themselves in the same outmoded manner.

Soon Would Develop a Costly Doctrine

From the misinterpreted events of 1918, U.S. war planners concluded that future expeditionary forces could project enough firepower to discourage any amount of frontal resistance. And that's how U.S. forces approached WWII.

Enough preparatory fire would be placed on each ground objective to make enemy countermeasures unlikely. But, such a plan ignored the roll of subsurface battlespace. U.S. forces could go wherever they wanted, but not without excess casualties. And no amount of bombing of "fleeing enemy soldiers" could justify this extra loss of U.S. life.

As Disturbing As the Thought May Be

Before the "War to End All Wars," U.S. troops had experienced the ambushing and short-range infiltration of 3GW during their many skirmishes with American Indians. But, the Germans' Spring Offensive of 1918 was their first exposure to mutually supporting strongpoints and "defense line penetrations through noisy surprise."

Those Doughboys had also dabbled in non-illuminated and shot-free night attacks. But, they mostly liked to realign, reorganize, and rest after dark. Then, in the Allied counterattack of 1918, they were forced to do everything the European way—through overly strict and fully predictable 2GW procedure. Keeping up with a rolling artillery barrage had been done at what price? Its associated assaults had already cost the British and French an entire generation.

The U.S. war planners' erroneous conclusion had come after fully exposed Doughboys had advanced against a receding band of under-supplied strongpoints in the Allied counter-offensive of 1918. Under tremendous pressure to justify the highest U.S. casualty count in the nation's history, those planners declared a 19th Century throwback to be standard procedure for any future opposition. In the process, they totally disregarded the Germans' tactical innovations of 1918. Only the bottom-up societies of America's Eastern adversaries would carry those 3GW techniques into WWII. That surprise and firepower were now interchangeable had been completely lost on the Americans. It was as if not showing GIs how properly to attack a machinegun would make its existence less deadly.

Along with this decision to go high-diddle-diddle up the middle on a regular basis would come fortifying the U.S. rifleman's psyche against the additional dangers. To the anti-European arrogance of 1917 was added the abject discipline of a Civil War assault line. But, instead of absorbing musket fire while stopping to reload, modern-day riflemen would advance against streams of long-range bullets. And binding this "2GW mindset" together would be unquestioning loyalty to one's organization and instant compliance with every order.

Had those post-WWI U.S. planners unwittingly doomed future GIs to the awful effects of standoff artillery and machinegun fire? Does 2GW's greater degree of control justify such a risk? Could any of WWII's casualties have been avoided? Those who have openly confronted an enemy automatic weapon would rather sneak up on the next one.[69]

The Supposedly Evolving Strongpoint Thread

During WWI, on-line tank/infantry assaults behind a rolling artillery barrage had made little exception for any intervening strongpoint. That bastion's automatic weapons would have to be quickly stormed, because the forward progress of the overall formation must be steady. Then, bypassed enemy gunners made this way of operating extremely dangerous.

11 World War II

- Was the machinegun threat remembered from WWI?
- Which armies kept marching into machinegun fire?

GIs assaulting foe machinegun across nearly level ground.

(Source: " Follow Me, by H. Charles McBarron, at Leyte, 1944," U.S. Army Center of Military History, artphoto archives, illustration designator "avop05-99_1.jpg,")

Control of the U.S. Armed Forces

At the start of WWII, the War Department had no operational control over American units. The President was their Commander-in-Chief, and professional soldiers planned all field maneuvers.

> At the beginning of World War II, the [War] Secretary's main duties were to determine general policy for the military establishment as a whole and to supervise the procurement of weapons, equipment, and supplies for the Army.[1]
> — National Archives, n.d.

That civilian Commander-in-Chief had delegated most of his operational authority through the various chains of command within the U.S. military establishment. The Joint Chiefs of Staff (JCS) were created only to enhance inter-service cooperation.

> The Joint Chiefs of Staff originated in World War II, when U.S. military leaders recognized the need for an overall command to coordinate the efforts of the various services. A unified high-command structure was adopted in 1942 and maintained on an informal basis throughout the war. The Joint Chiefs of Staff were formally established as an advisory body by the National Security Act of 1947, and a 1953 amendment to that act prohibited the body from exercising command authority.[2]
>
> — *Encyclopedia Britannica,* n.d.

For the duration of WWII, theater commanders exercised operational control over all expeditionary units. But then, from whom they got their initial guidance got a little confusing.

> [A] unified command over U.S. operational forces was adopted during World War II. . . . Unified command called for a single commander, responsible to the Joint Chiefs of Staff, assisted by a joint staff, and exercising command over all the units of his assigned force, regardless of Service. The system was generally applied during World War II in the conduct of individual operations and within geographic theater commands.[3]
>
> — History Office of the Joint Chiefs of Staff

How WWII would be fought remained the prerogative of the President and his military advisors. But, because of the vital role of new equipment, the civilian War Department became more and more involved—possibly to excess. For example, the limited shelf-life of ammunition may have led to a more firepower-dependent regimen.

Today, the U.S. military chain of command is quite different from what it was then. Administrative and operational control of the various Stateside and overseas commands tends to blur. But, the Secretary of Defense and appropriate Service Secretary are firmly ensconced within the chain of command for every military outfit.

Some believe the corresponding JCS member to be included.[4] But, with all that civilian influence, wartime decisions are now based upon more than just strategy, operations, and tactics. Because the economy stands something to gain, ordnance-expending attacks may far too often take the place those producing no collateral damage. One takes expensive firepower, while the other requires only the *gratis* element of surprise.

WWII's Misinterpreted Lessons

The American military would enter WWII without fully knowing what had led to its many casualties at the end of WWI. It honestly believed that enough armor and preparatory bombardment now permitted its infantry units to move forward against any degree of opposition.

The Yanks were proud of the difference they had made in Europe and had enjoyed all the homecoming adulation. But, within that fanfare had lain a subtle trap. Few changes seemed necessary to how U.S. grunts would operate in the next big conflict.

In the bitter clashes of 1917 and 1918, American bravado had seemed strategically helpful, albeit somewhat wasteful of manpower. It had supposedly intimidated the Germans during the final Allied push of WWI. But, those assaults had been against under-supplied and already retreating adversaries. Instantly closing with enemy militia would also work during a Banana War encounter. But, against well-dug-in and well-supplied German or Japanese regulars in the early 1940's, it was no longer a good idea. Nor were all those haphazard explosions beforehand. The top-down system had come to believe too much of its own budgetary exaggeration. Appearing to win a military engagement seemed the equivalent of actually doing so.

U.S. Marine Attempts at 3GW in WWII

After serving with the Chinese Communist 8th Route Army in 1937, Lt.Col. Evans Carlson formed the 2nd Marine Raider Battalion with its emphasis Maoist guerrilla effort. Lt.Col. "Red Mike" Edson had just followed the light-infantry model with 1st Raider Battalion. Both units were supposed to prepare foreign shorelines for U.S. invasion. But, something quite different happened. The 2nd would run a commando raid on Makin Island and deep indiction patrol on Guadalcanal, while the 1st blunted a massive assault against Henderson Field.[5]

Still, that original interest in limiting landing-beach fire had been on target. Early in the war, Gen. "Howling Mad" Smith had written a paper about disabling defender guns with pre-invasion commandos.[6] But by then, firepower had been so deeply ingrained in Corps thinking that any American presence (other than waterline frogmen) promised to mask U.S. fires and unduly risk American lives.

Four Raider battalions—two Maoist and two light infantry—were summarily "guillotined" (in the Raiders' own words) in early 1943. One battalion survived intact as an infantry outfit, but members of the other three were widely dispersed throughout the Fleet as machinegunners. Their bottom-up way of doing things had proven too disruptive for a top-down organization. Frontline common sense too greatly challenged headquarters control.

> Every . . . [Raider] would have the right to say what he thought, and their battle cry would be *"Gung Ho!"* [working together] . . . Officers would have no greater privileges than the men, and would lead by consensus [collective opinion] rather than rank.[7]
>
> — History and Museums Division, HQMC, 1995

But, Asia's bottom-up problem solving had predated Communism by eons. So, there was no reason to abandon it.

> Arrogate all authority to oneself but pool the wisdom of the masses. . . .
>
> Victory or defeat depends on the judicious employment of people. A shrewd general must be able to employ his people judiciously, and take full advantage of their talents.[8]
>
> — "The Wisdom of Military Leadership"
> *Six Strategies for War*

Troops from both sides during America's Civil War had elected company leaders, so some democracy in the ranks had been acceptable. Formal voting was seldom done, but an occasional show of hands broke no military protocol.

Military Wisdom Ignored Due to Foreign Origin?

The egalitarian processes of the Chinese Army had threatened Corps discipline. U.S. Marines were supposed to do as told with few questions asked. Their Chinese counterparts not only enjoyed more battlefield leeway, but also stood guard against commander exaggeration. "[Red Chinese] soldiers enjoy freedom of assembly and speech; . . . and . . . [after-action reports] are open to the inspection of all." The latter responsibility had nothing to do with Communism *per se*. By ancient Taoist tradition, Asians had always seen great dishonor in claiming to win a battle they didn't.[9]

In the West, many MW parameters are counter-intuitive—like attacking where the enemy isn't. But, within Asia, they make perfect sense. Luckily, a few ways of bypassing enemy resistance—the essence of 3GW—would survive the Raiders' disbandment. There would be decisive probing patrols at the following places: (1) up Suribachi, Hill 362-A, and Hill 382 on Iwo Jima;[10] (2) on the southern slope of Peleliu's Umurbrogol;[11] and (3) to Okinawa's Shuri Castle.[12] But "snooping and pooping" had been the Raiders' *forte,* not how safely to assault an automatic weapon.

U.S. Development of a Machinegun Position Take-Down

The Marines' initial bunker attack of WWII was the "blind 'em, burn 'em, and blast 'em" technique practiced on Tarawa in November 1943 (and re-enacted in the movie "Sands of Iwo Jima").[13] First, the bunker's gunport was subjected to small-arms fire, then a handheld flame thrower, and finally a satchel charge. But, two of the three steps involved running up to an automatic weapon. Just to get close to the Point gun on Peleliu in October 1944, two men had to crawl to one side through a smoke screen. In effect, the "blind 'em, burn 'em, and blast 'em" routine was made a little safer. White phosphorus grenades accomplished the first two steps, and a rifle grenade the third. Grenades were more plentiful than satchel charges, and thus became the weapon of choice for bunker take-downs. Satchel charges and flamethrowers were reserved for blockhouses.

But the Raider penetration of a New Georgia defense matrix in July 1943 had provided enough insight for Peleliu. When artillery and air support failed to materialize at Bairoko Harbor, the Maoist 4th Battalion Raiders had to rely on fireteam opportunism and a *gung ho* means of control. Their task would not be easy. Not only was the roof of each emplacement swept by machinegun fire from behind, but also preregistered for mortars. Those Raiders hadn't reconnoitered any part of that bunker belt,[14] or ever practiced assaulting mutually supporting positions. Nor did they have any flame throwers. So, they had little choice but to crawl forward on line and toss in explosives.

By all indications, some Raiders got through three of the four bunker strings with little more than "recon pull" and fireteam cooperation at Bairoko.[15] The unconventional 4th must have somehow outflanked every bunker in each row. That's most easily done by buddy team members crawling up opposing lane boundaries to double envelop its bunker. The 4th Raiders' direct descendent 2/4 would do something similar on Crescent Hill at Iwo Jima's crucial Sugar Loaf Complex in May 1945.[16] And the "horse mounting" that so intimidated the Japanese on Okinawa involved two GIs suddenly appearing above each cave opening.[17] The U.S. Army tells how such a thing may have been accomplished.

> They [the Americans] would bombard the Japanese positions they faced, forcing the IJA [Imperial Japanese Army] off the surface. They would then infiltrate men in small numbers through narrow gaps in the Japanese fixed-cave fire line ("dead spaces").[18]
>
> – *Leavenworth Papers No. 18,* by Huber

Such a maneuver was possible on Okinawa, because the forward faces of most U.S. objectives hadn't been fully protected. Their primary defense had been on reverse slopes.

By 1945, double envelopment had become an integral (albeit unauthorized) part of U.S. Army squad tactics. Sgt. Lewis' "special assault squad" used it at Item Pocket on Okinawa, and Army Rangers behind the Normandy beachhead.[19] On both occasions, a few squad members went right, others left, and still others to the center (to draw the foe's attention).[20] U.S. Army personnel may have also copied foe refinements to the same procedure. Flanking German elements used smoke to cross open ground and immediately advanced after shooting. That way, all return fire was directed at non-occupied spaces. The closest flanker then conducted the final assault with smoke and concussion grenades.[21]

But, there were few U.S. grunts worried about how much surprise might be lost through small-arms fire. Things that go "bang" or "boom" had so mesmerized them, that all "woods-sense" had vanished. Even concussion grenades were used for over-pressurizing a confined space.

America's Infantry Conference of 1946

The U.S. Infantry Conference of 1946 was concerned with two aspects of WWII: (1) too much dispersion of the action; and (2) too many U.S. units pinned down. It recommended that the squad's size and tactical doctrine be altered.[22] The new squad was to have nine men.[23] To lessen the number of "diversions," its internal structure wouldn't permit any cover fire for moving members. In fact, the new squad was never to have separate base-of-fire and maneuver elements.[24] The squad leader's greater degree of control would then better facilitate the parent unit's mission. But, also implied was a partial return to the unified squad management of WWI—wherein all squad members received the same instructions. The new Army squad would be less distracted by local circumstances, and every member more likely to do as told—namely, to keep advancing through enemy fire.

Why So Little Night Fighting?

Throughout WWII, U.S. units would seldom attack during periods of reduced visibility. The reasons are many. Enemy nighttime proficiency had forced them to realign at sundown to forestall infiltration and human waves. Plus, evening was a good tune for all the reorganization that heavy combat requires.

Some tacticians have blamed Western semantics for this aversion to night operations. The American definition of momentum was most easily achieved through preparatory shelling followed by an infantry assault. Whether that shelling had done any good was less apparent at night. Others saw the mechanics of a U.S. night attack as too complicated.[25] All that extra scouting and enemy line supervision would be easy to forgo at the end of a long day of heavy combat.

WWII in Perspective

A thinking adversary makes war less simple than it seems from an overstuffed chair. Decisive victory is hard to come by, and constructive criticism seldom welcome. But, for the sake of future U.S. riflemen, one must periodically look back into history to see if some new weapon or old trick might have make the going a little easier.

To field enough soldiers in a hurry in 1941, America had needed a military bureaucracy. But, as most civil servants will attest, tall top-down organizations can take on a life of their own. Their initial requirement for standardized procedures becomes a continuous struggle to update those procedures. In effect, the West's highly stratified bureaucratic model tends to resist any change, however beneficial.

That's when American troops encounter two kinds of opposition—one from the enemy, and the other from bureaucratic inertia. To overcome both at once, U.S. forces attempted a lot of end runs at places like Anzio in Italy and Kakazu Ridge on Okinawa. Some bypasses worked, and some didn't. But, almost all deception and maneuvering was done at the large-unit level, with composite elements still focused on the overall formation. Only enough subordinate-unit tactics were tolerated to preserve the forward motion of the regiment or division as a whole. And as in WWI, localized ground gains were often forfeited at the end of the day to better defend against counterattack.

At enemy defense lines in the Pacific, U.S. headquarters were often forced to let their frontline units do all the thinking.[26] That's the only way those units could keep from getting pinned down. The assumption from 1918 that enough firepower would allow that unit to keep moving had proven false.

Harder to Ignore Are the Limitations of Heavy Ordnance

In the Western Pacific, overwhelming firepower did little damage during a number of U.S. landings. Then, the U.S. Strategic Bombing Survey immediately after WWII carried with it the same unhappy message. It is in finally accepting this paradox of war that all future U.S. victories may lie.

At Peleliu in late September 1944, four hospital ships received 2,343 wounded Americans. The average life span for Marines in the first-wave of landing craft had been 46 seconds.[27] Then, it took 73 days of vicious, often hand-to-hand, fighting to take the island. The overall U.S. casualty rate—80% in some units—would the highest of any amphibious assault in the Pacific.[28] The preparatory bombardment had left something to be desired, and the most productive ordnance turned out to be a hand grenade.[29]

> Navy ships anchored offshore fired almost 6,000 tons of shells. Navy and marine aircraft dropped at least 800 tons of bombs. In their month on the island, the Marines expended almost 16 million rounds of ammunition, including 116,000 hand grenades.[30]
>
> — *Archeology Magazine,* July/August 2019

At Iwo Jima in February 1945, the landing beaches were undefended. Some big coastal-defense guns would prematurely reveal themselves during a run at the beach by reconnaissance party boats. Otherwise, the indirect-fire ambush of the landing zone would have proven more horrific.[31] But, no amount of preparatory bombardment would later pave the way for a ground advance.

> The process [final push after a day of rest] began at 7:00 A.M. of March 6, D-Day plus fifteen, in thundering barrages of artillery, naval gunfire, and wave after wave of carrier planes dropping clusters of napalm bombs as their machineguns and rockets ripped into enemy lines in low-level passes across the island. In tons of steel and explosives, the firepower approached that unleashed on D-Day in the final hour before Marines hit the beaches.
>
> Virtually every Marine howitzer on Iwo—132 75- and 155-millimeter guns from eleven artillery battalions—was in action. They first fired for thirty-one minutes on the western half of the island, and then rocked the eastern sector for the next thirty-six minutes. In sixty-seven minutes, 22,500 shells slammed down along the front, often as close as a hundred yards in front of the Marines waiting to jump off.
>
> From less than half a mile offshore, a battleship and three cruisers unloaded 450 rounds of eight- and fourteen-

> inch high explosives. Three destroyers and two landing craft moved close in along the western shoreline to plaster the cliffs and caves with mortars and rockets. . . .
>
> . . . [When the Marine units struck,] Japanese reaction was immediate and deadly across the island. Incredibly, the pulverizing pre-attack bombardment and air strikes seemed not to have affected the enemy at all.[32]
>
> — *Iwo Jima: Legacy of Valor,* by Ross

Most Japanese gun bunkers were hidden in such broken terrain that U.S. observers had to be within 100 yards to accurately call in artillery shells. Even when that could be done safely, the bunkers themselves defied destruction.

> [T]he organic [U.S. artillery] weapons were inadequate for . . . destroying the type of emplacements encountered. . . . Even the much heavier 155mm shells of corps units required ten [10] to 12 hits, all in the same place, to inflict major damages on many of the imposing Japanese fortifications (VAC ArtpRpt 29).[33]
>
> — *Iwo Jima: Amphibious Epic*
> Historical Branch, HQMC

Neither were the Okinawan beaches defended in April 1945. Kuribayashi had shifted the majority of his forces elsewhere and below ground.

> According to American military records, a total of forty-five thousand shells were fired from the big guns (over 5 centimeters) in preparation for the landing. In addition, [there had been] some thirty-three thousand rockets with an uncountable number of bombs dropped from the air. From the viewpoint of the defenders, it is exceedingly pleasurable to see this all wasted.[34]
>
> — *The Battle for Okinawa,*
> by battle veteran IJA Col. H. Yahara

America had planned to take Okinawa through mostly 2GW effort—with ineffectual bombardment as its weakest link. Such a paradox is hard to quantify during an unopposed landing. But, it became a mathematical certainty when three fully supported U.S. divisions tried to blow their way through the first enemy obstacles—the Machinato Line.

> The assault platoons advanced [on 19 April 1945], hopeful that the great mass of metal and explosive had destroyed the enemy or had left him so stunned that he would be helpless. They were soon disillusioned; for the Japanese, deep in their caves, had scarcely been touched, and at the right moment they manned their battle stations (XXIV Corps Actn Rpt, 37). Brig.Gen. Josef R. Sheetz, Commanding, General, XXIV Corps Artillery, later said he doubted that as many as 190 Japanese, or 1 for every 100 shells, had been killed by the morning artillery preparation (Appleman, *XXIV Corps History,* 179).[35]
>
> — *Okinawa: The Last Battle,* by Appleman et al
> U.S. Army in WWII Series

Then, in December 1945, came the final insult to American way of war. The *U.S. Strategic Bombing Survey* reported that German industrial production had actually increased during the Allied bombing campaign.[36] Full dispersal of any wartime asset can make it very hard to damage.

> None of the bombing did any good. German industry went underground and spread out.[37]
>
> — member of Strategic Bombing Survey Team
> "Last Days of World War II," *History Channel*

The Rolling Artillery Barrage Becomes a Thing of the Past

The U.S. Navy had conducted a rolling barrage during the Iwo Jima landing.[38] But, by the time the Battle of Okinawa occurred, the rolling barrage had mostly disappeared from America's play book. It couldn't be maintained at an intense enough rate to hurt well-dug-in defenders. At Biak, the U.S. Army had preceded its inland move with a heavy artillery barrage, but the cave-protected Japanese had been able to make a stand wherever they wanted and then counterattack.[39]

On Iwo, the rolling barrage may have helped the Marines to do four things: (1) quickly cut off Suribachi from the rest of the island; (2) escape the murderous artillery and machinegun ambush; (3) develop more individual movement technique; and (4) periodically bypass an enemy strongpoint.[40] But at what cost was all this accomplished? The U.S. casualty total for the first day would be over 2400.[41] Or, had the noise and spectacle of a rolling barrage been a morale builder, and way to keep U.S. troops moving forward through hard-to-hear yet still murderous fire?

On 21 June 1945 at the south end of Okinawa, the Leathernecks would chance another doctrine-violating double envelopment of an enemy concentration at Hill 72 on Kiyamu-Gusuku Ridge. Such a maneuver is good for taking prisoners, but doesn't permit much pre-assault bombardment of any kind.[42] With ultimate victory in sight, they had temporarily moved into the more compassionate 3GW arena.

How Much Help Was U.S. Armor to Yank Infantry in WWII?

Patton's tank-led breakout of the Normandy beachhead is a matter of public record. But, within the rugged terrain north of the Shuri Line on Okinawa, U.S. armor could only survive with close "ground pounder" protection. On that island's more open southern plains, it could range further afield with motorized infantry in tow.

> [H]and-thrown demolition charges . . . had been successful for the Japanese on both Saipan and Iwo Jima. But, although such desperate techniques often were successful on Okinawa, they almost always failed when [U.S.] infantry covering fire was present. The only time [American] armor attempted to operate without infantry support [prior to the Shuri Line], in the 27th Division's assault on Kakazu Ridge, the result was a disaster. [During the full 82-day struggle for the island,] Tenth Army suffered 153 tanks destroyed, with many more damaged. . . . In all, the casualty rate for American armored vehicles on Okinawa was high, approximately 60 percent.[43]
>
> – *Ryukyus: The U.S. Army Campaigns of WWII*
> Center of Military History Publication 72-35

And So WWII Ground Inexorably Onward

Hitting the Atlantic Wall had been so difficult that the AEF headquarters fully focused on it. Its units ha little prepared for the hedge row fighting to follow.[44] (See Figure 11.1.) When the U.S. pincers closed on a mostly empty salient at the Battle of the Bulge six months later, 50,000 enemy kills would be claimed from airstrikes along the 13 miles to the Hitler's Westwall.[45]

But, the Allies had arrived at a foolproof formula. Only three things were required to succeed in Northern Europe: (1) adequate numbers of personnel; (2) sufficient military materiel; and (3) a positive way of looking at things. At that infamous Battle of the Bulge, the AEF had placed an additional 600,000 troops into the path of the Germans.[46] Then, its commander went looking for more warm bodies.

Figure 11.1: U.S. Units Weren't Prepared for Hedgerow Fighting
(Source: "From Cornrow to Hedgerow," by Keith Rocco, National Guard Image Gallery, Historical Paintings, Heritage Series, ©.)

> [T]he Joint Chiefs of Staff . . . allocated . . . three more infantry divisions. . . . General Marshall began to comb out support units . . . [for] infantry replacements.
>
> What about transferring divisions from Italy, asked Eisenhower. Perhaps 100,000 U.S. Marines. He set his staff to work . . . obtaining volunteers from segregated Negro support units to join the infantry.[47]
>
> — *A Time for Trumpets,* by MacDonald

Such a formula wasn't heartless or dishonest, just the most logical way for a top-down society to conduct the fight of its life. Yet, such an operation would still prove costly with regard to friendly casualties and collateral damage. That's because the enemy's small units had been better prepared tactically.

> At the height of the battle for Bastogne . . . General Patton wrote in his diary. . . The Germans are colder and hungrier than we are, but they fight better." . . .
>
> I tried to interest the Pentagon in my . . . finding. . . . The typical answer . . . was, "Why study the German Army? We won the war didn't we?" Yes, . . . We won it because the Allies overwhelmed the Germans with numbers of men and machines.[48]
>
> — *Hitler's Last Gamble,* by Dupuy et al

> [T]he majority of American divisions needed overwhelming artillery . . . and close air support to best their German opponents.[49]
>
> —*For the Common Defense,* Millet and Maslowski

U.S. Way of War Had Only Progressed Along 2GW Lines

As WWII drew to an end, the Germans and Japanese both made their 3GW defense matrices more sophisticated. Their barbed wire had been mostly linear during WWI. But WWII produced bunker clusters surrounded by wire, with minefields and mortar concentrations in between. (See Figure 11.2.) Whenever a U.S. infantry unit entered a "gap," it risked annihilation from a firesack. Fake strongpoints also appeared that had been preregistered for a mortar barrage.

But, both U.S. infantry branches still tried to punch their way through those strongpoint belts—first hitting them with as much standoff firepower as they could, then heavy armor. Soon, the Japanese and Germans were hiding all defensive arrays from aerial observation, and concurrently expanding the subterranean battlespace.

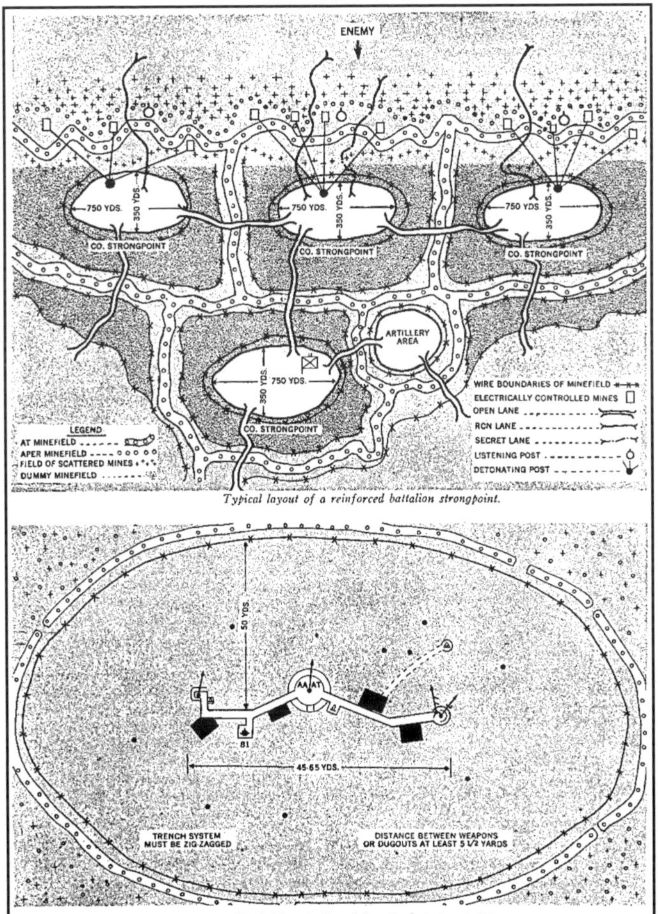

Typical layout of a reinforced battalion strongpoint.

Typical layout of a reinforced squad strongpoint.

Figure 11.2: German Squad Strongpoint Matrix of WWII

(Source: *Handbook on German Military Forces,* TM-E 30-451 (1945), p. 230)

U.S. forces responded by attacking in deeper formations to prevent spoiling attacks from their below-ground rear. Their infantry units still advanced on Iwo Jima, Peleliu, and Okinawa, but only because one or two American volunteers had crept forward to drop grenades into likely machinegun apertures from the back. Yet, the Second World War would end with U.S. headquarters and historians vehemently insisting that firepower had done the heavy lifting.

Veracity of This New Way of Looking at Things

The average reader may be wondering how comprehensive this "more objective" look into history really is. He knows his grandfather was proud of his WWII service. And, he has heard only good about the "Greatest Generation."

But, not to worry. There's good news, as well as bad. Those forefathers had been even more magnificent than the battlefield chronicles portray. Besides a very proficient enemy, they had been dealing with a lot of bureaucratic inertia (sometimes known as headquarters interference). Only their entry level of tactical proficiency had been exaggerated.

As WWII progressed, the U.S. military establishment kept attempting to organize what was—and always will be—a chaotic event. In this quest for order, it had increasing ignored mother nature, terrain abnormalities, enemy initiative, U.S. small-unit tactics, untested equipment, and any number of other factors. During the final battle for the Pacific, enough firepower would be "projected" ineffectively on Okinawa to rival everything non-atomic since then. American troops had been "danger close" to many of those explosions and suffered accordingly. In some ways, headquarters had unwittingly become as big a problem as the enemy.

Still, the Greatest Generation had achieved more than was legitimately possible with a 2GW agenda. What could have easily turned into contempt for authority was channeled into SNCO mentoring of young officers. The most crucial squad training was accomplished before any commissioned leader got to work. And the morale of U.S. infantry outfits skyrocketed. In essence, the NCO Corps of both infantry branches had assumed responsibility for the outcome of the war. Then, most headquarters lapses could be resolved through troop heroism, and all units kept moving. But, those brave young riflemen wouldn't make the ultimate sacrifice for public acclaim. They were demonstrating how properly to fight and win. Unfortunately, headquarters never saw it as such, and failed to adopt their tactical techniques. Luckily, the battalion commanders of this era were fine judges of small-unit potential. Some of the best had worked for Chesty Puller on Peleliu: Raymond G. Davis, Richard P. Ross, and Russell E. Honsowetz. Otherwise, the price for taking the most difficult terrain on earth would have been higher.

The Most Popular Ways of War Diverge

After brother killing brother in the "War Between the States," America had finally come up with a way to cross a battlefield with "a reasonable number" of casualties. The rolling artillery barrage was gone, but in its place had come enough "enemy fire suppression" to drown out the sound of incoming machinegun bullets.

So, most Americans (to include recent enlistees) thought the new WWII formula to be the optimum way of operating. All tactical lessons of the French and Indian and Revolutionary Wars were to be set aside. In their place would be a more modern system in which any forward element would mask one's fires. And because of continually improving hardware, advanced small-unit tactics were no longer needed.

But, the big Eastern armies hadn't arrived at the same conclusion. Instead of new paraphernalia to kill more people, they had looked for tactical ways—of any vintage—to counter the existing gadgetry. And, history would demonstrate which approach leads to more victory.

Not Much Progress in the Strongpoint Thread

Throughout WWII, U.S. infantry formations faced the same belts of tiny, mutually supporting machinegun nests as in WWI. But this time, there were mines, barbed wire, and firesacks in between the nests. To initially penetrate such a belt, a few advanced techniques appeared out of sheer desperation. Pairs of low-ranking volunteers crawled up lane boundaries to get behind enemy guns on Iwo Jima. And, once inside a belt, parent units attacked composite bastions from the side. But, those in severely broken terrain had to be reduced through semi-autonomous fireteam activity. When Iwo's narrow isthmus was crossed the first day, a few gun bunkers were bypassed, and fully exposed Marines quickly developed new moves. But, after the war, those ways of avoiding enemy fire were not considered important enough to become part of the normal training regimen.

Then, at Okinawa's Sugar Loaf Complex, alternating fireteam pushes up parallel lanes took the front face of Crescent Hill without any tanks, planes, or artillery. While somewhat like the piecemeal rushes of the Spanish American War, this Okinawan effort would be loosely controlled through fireteam cooperation *(gung ho* spirit). It had come from 2/4, the direct descendent of the Maoist-oriented 4th Raider Battalion.[50] Mao would also be the brainchild behind the human-wave simulating row of Stormtrooper squads in Korea and Vietnam. But, there is no evidence of "impromptu grinding" ever being added to any U.S. assault procedure. It would too greatly fly in the face of headquarters control. Only in small-unit "fire and movement" against a soft target would such a thing be occasionally permitted.

The U.S. system had been wise to under-encourage the frontal assault of any automatic weapon, but remiss in not showing every young grunt how to do so. As at Crescent Hill, he had to outflank the gun through microterrain. That would be, and still is, the state of the art for a gun bunker take-down. Of course, it would be better to bypass that strongpoint matrix altogether (and thus practice 3GW). But, with the onset of more powerful bunker-busting rockets, neither skill was thought as necessary. And, the impervious nature of Peleliu's Point fortification would soon be forgotten.

12 THE KOREAN WAR

- How intense was the infantry combat in Korea?
- Would the Pentagon learn anything from it?

Final segment of a U.S. hilltop assault.

(Source: "The Sunshine Division in Korea," by Rick Reeves, poss.©, from url: http://www.nationalguard.mil/resources/photo_gallery/heritage/images/sunshinedivision.jpg.)

Korea Starts Out the Same Way as WWII

To counter North Korea's invasion of South Korea in June of 1950, U.S. forces first defend the Pusan Perimeter and then land at Inchon in September. They take Seoul, but the Red Chinese intervene and a seesaw battle ensues. Operations Killer and Ripper are launched in February 1951 to retake Seoul. Heavily protected, "turtle-like" maneuver elements are able to project considerable firepower. Following the bloody hill battles and July armistice of 1953, the U.S. Army Infantry Conference recommends further changes to squad combat.

What Preceded the North Korean Invasion

South Korea had been in turmoil for several years before the war, and its president become quite dictatorial. Many citizens were suspected of Communist sympathies and killed on Cheju (or Jeju) Island off the south coast of the peninsula in the late 1940's. (See Map 12.1.)

> Starting in the spring of 1947, a group of Jeju islanders rose up against police brutality and called for a unified Korean government.
>
> The police and soldiers, joined by a right-wing paramilitary group from the mainland, responded with an extermination campaign, branding the insurgents as Communist agitators.[1]
>
> — *New York Times,* 28 May 2019

This was the same kind of terror campaign—complete with kangaroo courts and summary executions—with which Red Guards would operate during China's Cultural Revolution, and a People's Republic of China (PRC) proxy eliminate Soviet influence in Zimbabwe.[2] All Asian dictators knew how to conduct one.

> A vast majority of . . . [the protestors] were killed by police, soldiers and anti-Communist vigilantes hunting for leftist insurgents and their relatives.[3]
>
> — *New York Times,* 28 May 2019

But, there was another Cheju incident to shed light on how South Korean troops might treat everyone who failed to flee a North Korean invasion.

> Some 30,000 people were killed on Cheju. In October 1948, a [South Korean] constabulary unit at Yoshu refused to go there and mutinied.[4]
>
> — PBS Special, *Korea—the Unknown War*

> Though Kim Je-Hue—the Yoshu rebels battalion commander—had been my student while I was principal of the military academy, I volunteered to command the expedition to suppress the guerrillas in the South Chiri Mountains. It

Map 12.1: Country South of the 38th Parallel

(Source: Courtesy of "Nations Online Project," ©, for educational use from this url: http://www.nationsonline.org/oneworld/map/korea-south-administrative-map.htm.)

> hurt and broke my heart when I shot him and the whole battalion, but they were enemies that wanted our country to become communist and were therefore unforgivable, so I annihilated them before returning to Seoul.[5]
>
> — Chung Il-Wan
> South Korean Constabulary
> PBS Special, *Korea—the Unknown War*

Following the North Korean incursion of 1950, many of Syngman Rhee's soldiers would view all citizens who had failed to flee as Communist sympathizers.

> I took lots of pictures. Inchon was just a sea of rubble [after the landing]. . . . Now don't forget that all of these civilians are being liberated by the South Koreans. I think the South Korean soldiers were mad because they shot at everything that moved. And all these kids and old ladies are walking along with their hands up because they are terrified of these South Korean soldiers.[6]
>
> — Bert Hardy
> "Picture Post" photographer
> PBS Special, *Korea—the Unknown War*

To what extent this distrust of the civilian population in Korea would influence U.S, troops is not quite clear. Following rumors of Communist soldiers in mufti, columns of refugees were occasionally shot at to keep them from entering U.S. lines.[7]

> The thing that shook me more than anything was to see people—in extreme danger certainly—just shooting indiscriminately and even shooting civilians.[8]
>
> — Rt.Hon. George Younger M.P.
> Future British Defense Minister
> Then Argyll & Southern Highlander
> PBS Special, *Korea—the Unknown War*

How U.S. Marines Behaved After the Inchon Landing

The North Korean invaders had chosen to defend Seoul with "retreat combat," a 3GW variant in which only the actual routes of attack are fully contested to preserve limited assets. After encountering such a porous defense around Seoul, the Inchon Marines apparently threw all caution to the winds. Those fully exposed to enemy fire would pay a heavy price, but others did better. As at San Juan Hill in 1898, American machineguns with a well-defined target made it possible for some GIs to move forward.

> The 1st Marines commenced its main assault on Yongdungpo [a Seoul suburb] at 0630 the next morning. Neither Sutter's 2nd Battalion or Lieutenant Colonel Jack Hawkins' 1st Battalion could sustain much headway. Crossing the Kalchon was like crossing a medieval castle moat; clambering over the dikes was akin to "going over the top" in the trenches of World War I. Sutter's outfit in particular took heavy casualties. The division's Special Action Report recorded the loss of 17 officers and 200 men by the 2nd Battalion along the canal-like river by 21 September.
>
> Puller committed elements of Lieutenant Colonel Thomas L. Ridge's 3rd Battalion in the center, but a half dozen [North Korean Peoples' Army] NKPA Maxim heavy machine guns took a grim toll of every attempt to cross the water gate sector of the Kalchon.
>
> Ridge ordered Major Edwin H. Simmons, his Weapons Company commander, to suppress the fire. With his 81mm mortars temporarily out of ammunition and no artillery support immediately available, Simmons chose his Browning M1917A1 water-cooled .30-caliber heavy machine guns for the mission. Proven veterans of the World War, the heavy Brownings were unsurpassed in providing rock-steady, sustained fire at a rate of 450-600 rounds per minute. Simmons massed these weapons with their barrels "just clearing the top of the dike." A fierce duel ensued—"heavies against heavies"—at an interval no greater than half a football field. The exchange was deafening, but Simmons' sturdy Brownings prevailed, allowing 3rd Battalion, 1st Marines, to cross the Kalchon intact.[9]
>
> — History and Museums Division, HQMC, 2000

Then, either tremendous luck or Divine Providence took a hand. With little proof of any defender shortage ahead, a Marine commander followed a risky hunch. In row formation, he brought his entire company across a big rice paddy and stream of unknown depth. A pair of light machineguns could have killed every American. Even if that Captain had known only main avenues of advance are defended in 3GW, he was now the lead element on such an avenue. And there were North Korean snipers (doubling as artillery observers) strung out all across the city to ascertain where those invasion routes would occur.[10]

> The Kalchon proved a barrier to the entire regiment on 21 September—with one memorable exception. While the battle raged on both sides—and shortly before Major Simmons' machine gun duel—Captain Robert H. Barrow, the future 27th Commandant, led his Company A, 1st Marines, through a rice field towards an uncommonly quiet sector of the Yongdungpo defenses. The North Koreans may have vacated this sector in order to more effectively contest the adjacent water gate fronting the 3rd Battalion, an obvious crossing site. Barrow, however, expected to be hit at any moment. Simmons watched . . . as Company A, 1st Marines, advanced past his immediate left flank, each platoon on line. "They were beautifully deployed," said Simmons. "As they came through the dry rice paddy I thought of the Marines coming through the wheat fields at Belleau Wood in 1918."
>
> Private First Class Morgan Brainard of Barrow's company, though apprehensive about the spooky quiet, experienced similar thoughts as he crossed through the waist-high rice stalks. As he later described the advance:
>
> Somewhere off to our left, beyond the road and out of sight, beyond a line of trees we could hear the rattle of rifle and machine gun fire where Baker Company was going in . . . . To our immediate front, however, there was nothing but silence, as we continued to move forward through the field in perfect order. It was a classic-type infantry advance . . . but my mind kept racing back toward the stories I had read as a boy of the Marines attacking through the wheat fields of Belleau Wood . . . and I expected our peaceful scene would be shattered in a similar manner at any moment.
>
> "We passed over the top of the dike quickly, slithered down the other side," recalled Brainard, "then inexplicably and stupidly stopped facing a stream [the Kalchon]. I mean

> the whole line stopped." The company gunnery sergeant quickly ended their hesitation: "Get in that . . . water!"[11]
> — History and Museums Division, HQMC, 2000

But, opportunity had presented itself, and the Marine officer shown no fear of getting cut off and surrounded by greater numbers.

> Company A found itself entering the main street of Yongdungpo totally unopposed. "It was eerie," said Barrow. "We simply slithered into town undetected."
>
> The 87th NKPA Regiment, desperately attempting to patch together a defense in depth, had accidentally left this critical approach unguarded, and Barrow took full advantage of the opening. His 200-man company flowed rapidly into the heart of the city, sweeping up surprised bands of the enemy in the process. Before dark they had cut the city in two. Barrow selected a sausage-shaped dike, 30 feet high and 150 yards long, as the place to make a stand for the night. "We immediately recognized that we had a valuable piece of real estate," he said. From the dike his Marines could interdict the intersection of the highways from Inchon and Kimpo.
>
> Through this intersection at one point marched a large formation of unsuspecting NKPA infantry, singing political songs as they hurried to reinforce Yongdungpo's northwestern defenses. Barrow's interlocking machine guns and 60mm mortars cut down many and scattered the rest.[12]
> — History and Museums Division, HQMC, 2000

Operations Killer and Ripper

The Chinese People's Volunteer Army (CPVA) intervened in October and November of 1950, pushing U.S, forces well south of the 38th Parallel at several points across the peninsula. When the U.S. 8th Army counterattacked in February of 1951, it felt obligated to level everything in its path. American infantry units didn't match up well tactically with their Communist counterparts, so supporting arms picked up the slack. History records what happened as a type of meat grinder.[13]

> The advance of the seven American divisions now in the line was the twentieth century successor to the Roman "tortoise": instead of long columns, exposed to surprise attack, Ridgway's units now deployed at every stage for all-around defense in depth, securing themselves against infiltration while they waited for the massed artillery and air strikes to do their work upon the Chinese positions. On March 7 Killer was succeeded by Ripper. . . . The envelopment of Seoul by the success of Ripper made the Communist evacuation of the capital inevitable. On March 14 the victors recovered a devastated city, a metropolis of ruins and corrugated iron in which, of the principal buildings, only the Capitol and the railway station survived.[14]

Such a meat grinder would likely prove fatal for any nearby civilian, however well-burrowed.

> 40 million gallons of napalm were used in Korea.[15]
>
> — PBS Special, *Korea—the Unknown War*

To the South Korean residents being liberated, Ridgway's stratagem seemed terrifying. Anyone who took cover in a hole ran the risk of being later confused with an enemy soldier. What were mostly peasants had experienced invasion before, but nothing involving this much personal danger.

> The North Koreans were like ghosts. They passed over the countryside and left no mark on it in many ways. But when you use the rock crusher techniques of an American army you hurt your friends. And that was true in Vietnam as well as in Korea.[16]
>
> — Maj.Gen. Edwin Simmons USMC (Ret.)
> former Director of USMC History & Museums
> PBS Special, *Korea—the Unknown War*

Throughout the Korean War, the Chinese and North Koreans had little trouble infiltrating Allied lines. (See Figure 12.1.) Both Communist armies had harnessed the common sense of each soldier by decentralizing control, and his tactics resembled that of a guerrilla. With the distinction between guerrilla and civilian starting

Figure 12.1: U.S. Lines Were Often Infiltrated in Korea
(Source: "Corporal Hiroshi N.Miyamura, Korean War," by George Akimoto, U.S.Army Center for Military History, from this url: http://www.history.army.mil/art/A&I/0507-3.jpg.)

to blur, more than one Allied unit paid too little attention to the indigenous population. South Korean civilians in the path of Operations Killer and Ripper would have a very hard time.

DoD Had Grown More Powerful Since 1947

The relationship between the DoD's civilian leaders and the Joint Chiefs of Staff had begun to waiver a little.

> U.S. Department of Defense . . . [is] responsible for ensuring national security and supervising U.S. military forces. Based in the Pentagon, it includes the Joint Chiefs of Staff, the departments of the U.S. Army, U.S. Navy, and U.S. Air Force, and numerous defense agencies and allied services. . . .
>
> . . . The original National Security Act of 1947 provided the secretary of defense with limited authority and staff to exercise control. However, subsequent amend-

> ments to the basic act in 1949, 1953, and 1958 greatly strengthened the authority and ability of the secretary to direct defense policy. . . .
>
> The civilian secretaries of the military departments (Army, Navy, and Air Force) are responsible for the administration of their particular departments and are responsible for training, equipping, and providing combat-ready forces to the combat commands. . . .
>
> The commanders of the unified commands are responsible to the president and the secretary of defense. By secretarial delegation, the Joint Chiefs of Staff exercise operational direction over the unified commands.[17]
>
> — *Encyclopedia Britannica*, n.d.

Not until 1953 would those high-level relationships get sorted out.

> The Joint Chiefs of Staff were formally established as an advisory body by the National Security Act of 1947, and a 1953 amendment to that act prohibited the body from exercising command authority.[18]
>
> — *Encyclopedia Britannica,* n.d.

The Asians Had Become Quite Adept at 3GW

One the best-loved generals of WWII was Chesty Puller. Like Patton, he great respected the lowly Private and constantly strove to help him. Such help often took the form of "further discussions" on frontline realities with higher headquarters.

> I'm afraid we haven't recognized the most important lesson from Korea. The Communists have developed a totally new kind of warfare. . . . This is a total warfare, yet small in scope, and it's designed to neutralize our big . . . weapons. Look at Vietnam. The French outnumbered the Communists two to one, yet they [the French] were massacred.[19]
>
> — Chesty Puller

> My prayer now is that our leaders, knowing that we have no war machine, will evacuate Korea completely, have a thor-

ough house cleaning, and then rebuild a real war machine before becoming involved in another war. May God give us wisdom and common sense! [20]

— Chesty Puller
when sent home from Korea and retired

Might Chesty have been talking about the lack of bottom-echelon maneuver that top-down bureaucracies try to replace with centrally controlled bombing and shelling?

Firepower Had Once Again Under-Achieved

During the Korean War, there were several occasions in which massive U.S. bombardment had failed to meet expectations. As U.S. Army occupation troops were forced back into the Pusan Perimeter, in the summer of 1950, they suffered 19,759 casualties: 4,599 killed, 12,058 wounded, 2,701 missing, and 401 captured. Then, when they had to breakout of a Chinese encirclement at the Chosin Reservoir in late November 1950, they took another 17,843 casualties: 1,029 killed, 4,894 missing, 4,582 wounded, and 7,338 non-battle losses.[21]

But, the classic example of how little airstrikes, naval gunfire, artillery, and armor can sometimes help a U.S. ground operation came in 1953. That's when a fully supported United Nations (U.N.) phalanx attempted to penetrate North Korea for a second time and was unable to do so. The reason, of course, was a new Communist defense zone with overlapping strongpoint belts and "localized" supply depots, all below ground.

During that time [just preceding the 1953 Korean hill battles], anyone who happened to enter the mountainous area of our positions could hear the sound of earthwork. . . . Gradually, a defense system took shape. It was backed up by supporting [strong] points with tunnel fortifications as main structures. . . . Here one could find all kinds of facilities . . . as well as defended fortifications, communications entrenchments, and crisscrossing main tunnels and branch lines. . . . Even though the enemy continued dropping thousands of bombs that exploded on top of the mountains, our

> commanders were able to sit down and peacefully read books and newspapers or play cards and chess.[22]
>
> – *Mao's Generals Remember Korea*

Most Hill Battles Occurred in That Defense Zone

The CPVA had helped the North Koreans to build a very sophisticated "defense in depth" just north of the 38th parallel. No amount of U.S. mechanized firepower could fully penetrate it.

> From the Sea of Japan, on the east, to the Yellow Sea on the west, they burrowed into the earth. They entered mountains from the rear slope, tunneled through to make gun positions opening on the front. . . . They dug bunkers in which a company . . . could bivouac. They dug so deeply into the earth that no conventional gun or cannon could reach them.[23]
>
> – *This Kind of War,* by Fehrenbach

And that tunneling consisted of more than just an entrance passageway to each strongpoint cluster. As on Iwo Jima in 1945,[24] most clusters were linked below ground.

> The total length of the tunnels dug by our army was 1,250 kilometers. [Near the 38th parallel, the Korean Peninsula is only 193 kilometers wide.] [25]
>
> – *Mao's Generals Remember Korea*

> The CPVF dug more than 780 miles of underground tunnels, and the various kinds of surface entrenchments and communication lines along the front that amounted to . . . 3,900 miles. . . . CPVF Command . . . [was] to standardize the construction. . . . [E]very underground tunnel should be at least 100 feet deep . . . , have more than two exits, . . . and be connected to trenches or other fieldworks.[26]
>
> – Chinese museum records
> in *China's Battle for Korea,* by Xiaobing Li

The Communists hadn't built parallel defense belts along a gently curving trace, but to coincide with North Korea's random patchwork of ridgelines. To partially give way under heavy pressure, this barrier consisted of a series of "soft" belts of sub-surface (yet still mutually supporting) strongpoint matrices. That's why at Bloody Ridge—the first U.N. objective after the start of the truce talks—foe defenders eventually pulled back to Heartbreak Ridge.[27]

> [After building the first below-ground strongpoint array,] they [the Reds] went backward and dug a new defensive line, and one beyond that, stretching to the north. They dug a . . . [defense zone] such as the world had never seen—ten times the depth of any in World War I (WWI).[28]
>
> — *This Kind of War,* by Fehrenbach

Of course, the Communists had also built all the surface bunkers, covered trenches, and open firing steps necessary to protect such an underground labyrinth.[29]

> The various kinds of entrenchments and communication lines that we dug amounted to the total length of 6,240 kilometers. . . . [T]he earthwork construction of more than 100,000 bunkers.[30]
>
> — *Mao's Generals Remember Korea*

In all, three belts of permanent defense works had been constructed. Exactly where has been openly detailed in *China's Battle for Korea,* by Xiaobing Li.[31]

Most engagements of this period occurred inside this defense zone. At the battle of the Punchbowl, one American GI noticed something that was probably true of others near by.

> Hill 749 proved to be a heavily defended fortress of bunkers, covered trenches and tunnels and part of the [N]KPA MLR [Main Line of Resistance].[32]
>
> — Punchbowl veteran in his Masters Thesis

Most Hill Battles had been preceded by a massive U.N. bombardment. No fewer than 451,957 artillery rounds had been fired at Bloody Ridge alone, and then another 229,724 at Heartbreak Ridge.[33] Yet, the enemy's defenses still held.

> The [Red] bunkers were fortified to withstand air and artillery pounding. . . . Dug into the rubble of partially wooded

> slopes, obscured by morning mists, the North Korean positions were almost impossible to detect.[34]
>
> — *This Kind of War,* by Fehrenbach

One of the last of struggles of the war (and also the farthest north) occurred in July of 1953. It was for the Kumsong River Salient—an outpost occupied by South Korean troops. After the Reds took back the salient, an all-out U.N. counterattack was unable to dislodge the newcomers.[35] All fighting soon ended with the signing of the Korean Armistice Agreement.

Amount of Civilian Death in North Korea

The U.S. military may have countered Communist aggression in Korea, but it also killed many civilians in its bombing of the North. As in WWII, that bombardment would have little effect on the enemy's resupply effort.

> The terrible irony of the bombing of N. Korea was that it didn't affect the war effort. . . . We never found a way to stop their supply lines for example. These little guys with A-frames on their backs carrying several hundred pounds of supplies could not be hit by bombing. The reason for this is very simple. It goes back to the bombing survey done of WWII. . . . Bombing is not an effective strategy for ending a war or stopping supply lines.[36]
>
> — Professor Bruce Cumings
> PBS Special, *Korea—the Unknown War*

> 600,000 tons of bombs fell on N. Korea. Possibly 2 million civilians perished.[37]
>
> — PBS Special, *Korea—the Unknown War*

Reds Didn't Want U.S. to Change Its Outmoded Way of War

At the end of WWII, heavy preparatory fire had been of little help to U.S. maneuver elements on Okinawa.[38] But, the enemy chose a heavy shelling form of 3GW to keep the Americans from moving beyond this 2GW mindset. The Japanese even announced attrition as their way of keeping the Americans away from the home islands.[39] But, their style of war remained decidedly 3GW. Every attack was by infiltration, and *banzais* a thing of the past. Every hilltop was defended by ambush from its reverse slope. All the while, U. S. bombing and shelling continued with its fair share of short rounds. And, not only U.S. forces, but also indigenous civilians paid a terrible price for the island's liberation.

In Korea, the enemy would use the continuous threat of human-wave assault to keep the Americans reliant on their old-fashioned linear defense (with its FPL of interlocking machinegun fire). Each series of human waves at the Chosin Reservoir was actually a row of stormtrooper squads in extended column. Upon reaching the U.S. wire, those squads took turns attacking in a much safer grinding motion.[40] But, U.S. forces never realized the deception. Nor did they shift to the more productive 3GW (or strongpoint belt) way of defending.

On both occasions, the foe hadn't wanted the Yanks tactically evolving at the small-unit level. They knew that their own 3GW agenda would eventually trump America's 2GW procedures. And they were counting on her top-down bureaucracy and bottom-echelon loyalty to perpetuate a less-than-adequate way of war.

Infantry Conferences Associated with Korea

U.S. Army squads had entered the Korean War without any internal fire and movement capabilities. But, near its end, their parent organization would alter squad dynamics. Despite the conclusions of the 1946 Conference and other studies, Army leaders increased the squad's size from nine men to 11 and returned to the 1940's doctrine of fire-covered maneuvers within.[41] B.Gen. S.L.A. Marshall had seen squad members forced apart by Korea's severely broken terrain, making them more difficult to control. After he recommended the squad no longer function as a unified entity, the TOE was changed in 1953. This enabled each 11-man group to subdivide into two equally balanced fireteams (each with its own [Browning Automatic Rifle] BAR). Because the rugged landscape caused so much dispersion, many parent units had already experimented with two teams. In the attack, each rifle squad would now have its own base-of-fire and maneuver elements.[42]

Not until years later were all changes formally adopted, but the Army hadn't based this decision upon preventing casualties (the acknowledged indicator of good small-unit tactics). Among the key factors on squad sustainability at the 1946 Infantry Conference had been: (1) maintaining adequate levels of ammunition; and (2) absorbing [sufficient] casualties.[43]

And as had happened with the 1946 conference, the core issue in 1953 was how to keep the infantry squad moving through enemy fire. The combat experiences of M.Gen. Fry, commander of 2nd Infantry Division in Korea had produced little empathy for the frontline fighter.

> [M.Gen.] Fry claimed that his introduction of fire and maneuver teams eliminated "pin-downers," soldiers who could not move because they were pinned down by enemy fire.[44]
> — *Combat Forces Journal,* April 1953

In other words, the "big picture" and Western version of momentum would continue to subject enlistees to unsuppressed fire. Within "bottom-up" Asia, momentum is not achieved through forward motion, but rather consecutive small-unit victories involving as much surprise (and little danger) as possible.[45] So, which side of the world has been best limiting friendly casualties?

The Strongpoint Thread Shows Little Progress

Any North Korean strongpoint in the way during America's retaking of Seoul in 1951 was overwhelmed through a large, tortoise-like maneuver element packing a lot of firepower.[46] There was no attempt to bypass that strongpoint or outflank its machinegun bunkers. And, GIs within that formation still suffered considerable casualties from long-range fire. But, when the Reds assaulted U.S. lines at the Chosin Reservoir, they did so through a row of semi-autonomous stormtrooper squads in column. What had initially looked like a series of human waves then started a random grinding away of U.S. lines.[47] Which of the two maneuvers would be more respectful of human life?

13 The Vietnam War

- Was much enemy envelopment practiced in Vietnam?
- Why not?

Little time existed for hole searches during large-unit sweep.

(Source: "Perimeter Patrol," by SP4 Michael Crook, U.S. Army Center of Military History, posters, illustration designator "p_3_4_67.jpg,")

Something Easier to Understand in Retrospect

From 1965 to 1973, U.S. Army infantry squads in Southeast Asia still followed the "fire and maneuver doctrine" from 1953. The now 10-man unit contained a squad leader, two four-man fireteams, and an M-79 man or radio operator.

Near war's end, the U.S. Army would increase its rifle squad size to 11. Its Infantry Conferences had concluded the abundance of small-unit action caused too many casualties for proper squad functioning. And the previous structure didn't allow enough fire and maneuver within each squad.[1]

But, the need for tactical reform had only been insinuated. Better fire and maneuver internally would take less shooting and more moving.

U.S. Forces Fight Bravely But Take Many Losses

While the threat of human-wave assaults still existed in Vietnam, the foe more often conducted 3GW infiltration. Silent sappers went after command posts, mortar pits, big guns, and ammo dumps, as lone stormtrooper squads attempted to enter U.S. perimeters by noisy surprise.

Still reliant on 2GW defensive lines, U.S. forces best countered this enemy effort through the bombing and shelling of withdrawing attackers. And, a certain amount of preparatory fire also preceded every American assault. In Vietnam, the unfortunate side effect of this old-fashioned way of fighting was fully quantified by an official study.

> As our [U.S. Army] study shook out, the fact became inescapable that a staggering 15 to 20 percent of all U.S. casualties in Vietnam were caused by friendly fire.[2]
>
> — *About Face,* by Col. Hackworth

How the Americans Almost Succeeded

A Marine infantry battalion discovered early in the war what it would take to win this conflict.

> Combined Action . . . started in August 1965 as a unit drawn from 3rd Battalion, 4th Marines . . . in the Phu Bai area. 3/4's [Tactical Area of Responsibility] TAOR included six villages. . . . [Its] executive officer suggested that they incorporate local militia units into 3/4's operations. . . . [Plan forwarded] to III Marine Amphibious Force (III MAF). . . . General Lew Walt . . . agreed to the proposal. . . . [3/4] integrated four [Marine] squads with the local PF [popular force] units in August 1965.[3]
>
> — *Our War Was Different,* by Hemingway

Gen. V.H. "Brute" Krulak at Fleet Marine Forces Pacific (FMFPac) was highly supportive after Maj.Gen. Walt, a former Marine Raider, fully implemented the Combined Action Platoon (CAP) program. Having been with 4th Marines in China from 1937 to 1939, Gen. Krulak was well aware of the "Spreading Inkblot Theory," in which government control spreads through a village presence.[4] This was also Mao's guerrilla strategy, so the FMF Head wanted to counter it through similar activity. Both U.S. generals had learned from Asian experience that the armies of bottom-up societies are most easily beaten from the bottom up. Walt's below mention of "passing through" is a direct reference to the ineffectiveness of large-unit sweeps.

> "The struggle was in the rice paddies . . . in and among the people, not passing through, but living among them, night and day . . . and joining with them in steps toward a better life, long overdue."[5]
>
> — Maj.Gen. Lewis W. Walt USMC

From 1965 to 1971, some 5,000 Marines and Navy Corpsmen were to operate virtually alone in isolated villages from Chu Lai to the Demilitarized Zone (DMZ) as part of the CAP program. Though ostensibly pacified, most of those villages were thoroughly surrounded by difficult terrain, booby traps, and hostile elements. Each had a joint platoon comprised of a Marine squad (with its Navy Corpsman) and 30 to 35 PFs. By 1969, the number of platoons had reached 102. They were organized into 19 companies under four Combined Action Groups—the 1st through the 4th.[6] (See Map 13.1.) Most U.S. squads had been winnowed away from a nearby infantry battalion.

A 2/4 inspector describes the role of the early CAP unit in this way. "Their mission was to protect the village against Viet Cong (VC) infiltrators and to assist in the general pacification programs." The latter included everything from medical and construction aid to protection against enemy confiscation of food and conscripts. The CAPs around Phu Bai had few, if any, preplanned artillery concentrations to keep from being overrun.[7]

Some roving CAPs also existed. They moved back and forth between villages after spending every night in the field. Their mobility kept the enemy guessing as to their exact whereabouts. They would set up temporary command posts (CPs) at various locations

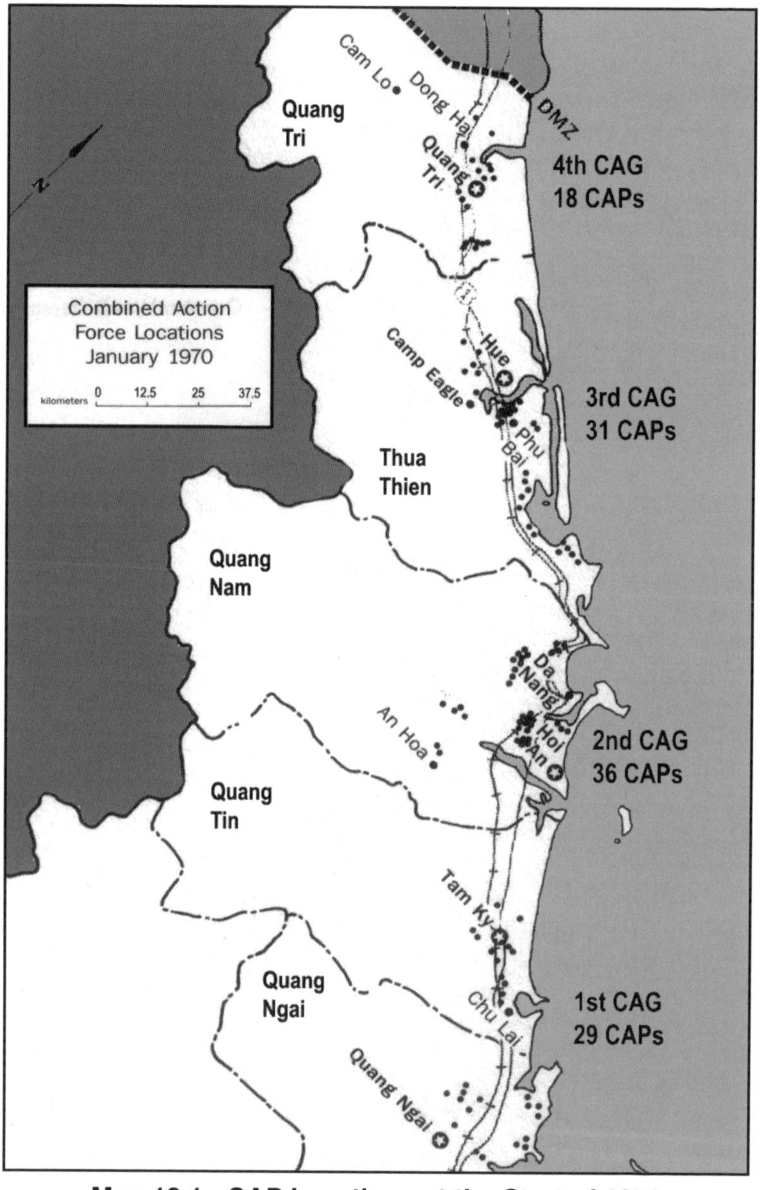

Map 13.1: CAP Locations at the Start of 1970

(Source: Map based on USMC Hist. Ctr. drawing by W. Stephen Hill, from "Our War Was Different," by Al Hemingway, © 1994 by Naval Inst. Press) throughout a TAOR, and there rest during the day (just as the NVA regularly did at paddy islets). At night, all CAP CPs were manned by a single fire team. Everyone else was out on a patrol or ambush.[8] But, such an imaginative way of handling Communists was not to last. By the end of 1968, the U.S. president had abandoned his bombing of North Vietnam to more widely apply Gen. Westmoreland's "search and destroy" method.

> [Pres.] Johnson chose William Westmoreland . . . to command . . . [Military Assistance Command Vietnam] MACV in June 1964. Over the next four years, the general directed much of U.S. military strategy. . . . His strategy of attrition aimed to inflict heavy losses on North Vietnamese and Viet Cong forces using superior U.S. firepower.[9]
>
> —"Westmoreland," *History Channel,* 29 May 2019

> In pursuit of his strategy of attrition, [Gen.] Westmoreland . . . was seeking to bring the total number of troops up to 550,500.[10]
>
> —"Westmoreland," *History Channel,* 29 May 2019

> In September 1967, when (foe) . . . forces began . . . attacks on . . . Khe Sanh. . . . Westmoreland saw this as a positive development. . . . After [Allied] . . . forces inflicted heavy losses . . . Westmoreland reported to Johnson that the end of the war was in sight, as the Communists could not possibly replace the men they had lost.[11]
>
> —"Westmoreland," *History Channel,* 29 May 2019

By the end of the Johnson administration, the number of U.S. troops in country would closely approximate Westmoreland's number. And the president's foreign policy goals also matched Westmoreland's reason for the heavily armed sweeps—more attrition than the enemy could absorb.

> The number would surge to 535,000 by the end of Johnson's presidency. . . . The President's "middle way" involved a commitment of U.S. ground forces, designed to convince the regime in Hanoi that it could not win.[12]
>
> — Univ. of Virginia Center on Presidential Policy

Shortly after President Nixon's inauguration, it was decided to turn the war over to the South Vietnamese. This made the CAP initiative no longer necessary. Its brilliance would not be realized for many years.

Within the tactical details of CAP program lay its strength. "Tactical mobility" had been assigned to stationary platoons. The outskirts sentry array of each village was in motion. It mainly consisted of a screen of shifting ambushes.

> A CAP does not defend . . . from behind bunkers and barricades. The idea is not to put up a wall around the hamlet, but rather to put out a screen of ambushes. . . . When coupled with stealth, [such] mobility provides . . . the protection afforded by elusiveness. . . . [T]he CAP seems to be everywhere, but never predictably anywhere. . . . [This] is the basis for CAP security against surprise attack by overwhelming enemy forces.[13]
>
> — CAP "Fact Sheet," III MAF, 1970

Such protection gave each U.S. contingent a good chance of survival. By August 1967, there were 75 CAPs and another 39 authorized. Only one or two had been compromised in two years of the program's existence. And then, there would be only 11 more compounds penetrated over the next four years.[14] By end of 1967, 25 fully pacified villages were ready for turnover to their PFs. When CAP platoon totals peaked in 1970, 93 Marine squads had been shifted away from villages capable of defending themselves. "Of the 209 villages protected by CAP units, not one reverted to VC control." To this day, that remains an unparalleled record of combat efficiency.[15]

But, the Pentagon would persist with its lackluster plan—conventional battalions continually sweeping the area in search of enemy forces. In a 1965 statement to reporters, even General William Westmoreland had pointed out the plan's inherent weakness. With such a strategy, the Vietnamese peasant would have three choices: (1) stay close to his land [in a free-fire zone]; (2) join the Viet Cong [the U.S. target in that zone]; or (3) move to an area under South Vietnamese government control [become a refugee].[16] So, for or all practical purposes, the Pentagon had declared war on the country's peasant population, just as the British with Malaya's rural Chinese in the early 1950's.[17] But, unlike the Malay Peninsula, South Vietnam had a lengthy mainland frontier across which guerrillas could be easily reinforced and resupplied. So, what had marginally worked in Malaya would have little chance in Vietnam.

The Real Problem with Combat Sweeps

America's search and destroy formations were an attempt to maximize combat power by keeping all advancing combatants and their weapons on line. No opposition was then supposed to stand a chance. But, what if that opposition were hiding below ground? Then, the all-powerful U.S. force could unwittingly pass over strongpoint arrays, and its members get shot in the back as on Iwo Jima. Unless that formation continued onward, it would violate the U.S. definition of momentum. So, any automatic weapon had to be charged, and suspicious holes left unsearched.

From high overhead or back at headquarters, the whole affair looked quite orderly. Enough bodies of possible enemy combatants were discovered to make it seem worthwhile. And, the war planners stuck to their inane agenda. But, actually vanquishing a bottom-up adversary takes more attention to detail than that.

To Decisively Win a War in Southeast Asia

U.S. veterans of Vietnam often say their hands were tied. But a certain amount of restraint helps to preserve one's humanity, so most vets aren't talking about looser rules of engagement. Some wanted to invade North Vietnam, while others would have settled for "bombing Hanoi back into the Stone Age." But neither approach would have worked. Any attempt to directly cross into North Vietnam would have been met with disaster.

When the Chinese ran a fully supported punitive expedition into North Vietnam to stop its Cambodian meddling in 1979, they took horrific losses. The only barrier had been a sub-surface strongpoint belt of "military fortress" design. (See Figure 13.1.) And, this border-long array had been manned by militiamen. They had mortars, antitank guns, and machineguns, but no tanks, planes, or artillery. Most of their NVA brethren had been down in Cambodia contesting the Khmer Rouge. After being cut to pieces by all the interlocking bands of fire, the Chinese did manage to take a few border towns. But, they quickly declared victory and withdrew before the North Vietnamese regulars could return.

In effect, the Chinese had been tempted by technology to try the U.S. way of war. But in the South, the North Vietnamese had already discovered how to defeat a U.S. phalanx without any of its own planes, tanks, or artillery. Nor did they need much resupply. Everything required could be stolen from the oversupplied foe, or smuggled in. That's why no amount of bombing of Hanoi and

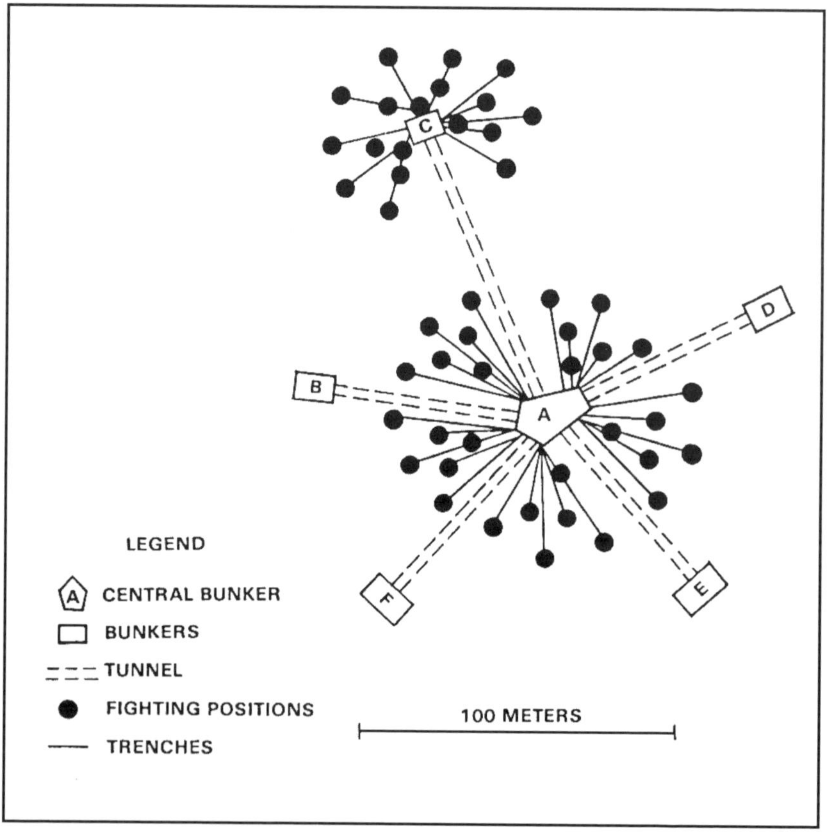

Figure 13.1: Possible Vietnamese "Military-Fortress" Segment
(Source: *Counterguerrilla Operations,* FM 90-8 (1986), p. A-6)

Haiphong would have brought North Vietnam to its knees. When pitted against the Pentagon's strictly 2GW agenda, the NVA's knowledge of 3GW would have eventually produced a Communist victory.

Some Casualties Are Impossible to Avoid with 2GW

The way most U.S. infantry companies fought in Vietnam was sure to produce friendly casualties. Much of the war was waged through cordon operations and on-line sweeps. Though each sweep was methodical, it moved too fast for participants to pay much attention to the ground beneath their feet. While the least experienced tripped booby traps, others got too close to an underground chamber and sniped. Trying to keep the troops alive became a real chore for commanders. It was hard for a U.S. rifleman to last six months, much less a full 13-month tour.

This seemingly unavoidable bloodletting, in turn, caused most maneuver elements to become overly defensive. While sitting atop a highly contested hill southwest of Da Nang in October of 1968, U.S. Marines of G/2/5 wanted to form killer teams to counter the NVA's nighttime probing. The Commanding Officer (CO) told them it was too risky.[18] Not only was moving around just outside a Marine perimeter liable to draw friendly fire, but the foes in this region were expert light infantrymen. As logical as that commander's decision may now seem, he and the whole system were making the classic mistake against a "bottom-up" adversary. The insights of a "mobile" CAP defense were not being followed. Too much caution had caused this unit to be over-controlled. With little, if any 3GW training, its CO had thought the troops too tired or hastily trained to reverse their fortunes. Such an attitude on Iwo Jima would have cost America the war. There wouldn't have been any low-ranking volunteers crawling forward to silence machineguns. Any unit morale problem is a direct reflection on the commander's leadership ability.

To this day, the average young American grunt is unbelievably strong, brave, smart, and in possession of unlimited growth potential. No one knows exactly why. It may come from his home environment, street experience, good genes, or military training. But, from wherever it derives, it's almost always there. That means the Pride of Kokomo can accomplish almost anything when not babied or micro-managed. An over-emphasis on safety is not good for him. His wartime ability must be allowed to incrementally grow. Then, at some point, he becomes more than a match for any Eastern counterpart.

What It Was Like in the Vietnamese Field

When strategy drives operations and operations dictate tactics (as they do with U.S. forces), headquarters is continually plagued with G-2 "discoveries." Then, to maintain the Western version of momentum, it creates too high a tempo of operations. While resting at an east-west trail junction near the DMZ in late 1966, A/1/4 experienced the all-too-common "missed opportunity." A platoon-sized patrol discovered the grass temporarily crushed down at one meter intervals along a trail to the north. Upon asking permission to investigate, the platoon leader was told there were helicopters inbound.[19] There's no telling what a row of crawling Leathernecks might have uncovered that afternoon with enough time to fully develop a combat situation.

The infantry battle in Vietnam was largely fought with supporting arms. Most battalion TAOR's had been designated "free fire zones," and the locals asked to leave. Then, whenever resistance occurred *en route* to a G-2 hot spot, an artillery or air strike was called in. Periodic sniping from a village didn't qualify, but everything near the objective did. If there were no supporting arms available, a frontal assault was run. Only where a rice paddy intervened was any other maneuver even contemplated. That frontal assault may have had a base of fire off to one side, but little surprise was possible There was no standing order to operate this way, just how things were done. And it only slightly varied between battalions. In effect, looking good at the top had taken the place of being good at the bottom.

Few things were fully investigated during any sweep. That would have too greatly affected its direction and speed. So, every antagonist got the chance to escape. Most could have hidden below ground until the Marines left. They had chambers with more than one exit—some below the surface of a tidal stream. Only after a battle would Marines stop to look for buried contraband. Otherwise, they just kept crisscrossing their TAOR for 28 out of every 30 days, ostensibly to control it. (See Figure 13.2.)

The Strongpoint Thread Makes No Headway

Most enemy strongpoints were carefully camouflaged and only partially protruding in Vietnam. So, U.S. sweeps regularly passed right over them. Nor was there any special provision for the danger they posed. Whenever a hidden bastion chose to fight, sweep participants had to improvise. At the Battle of Dai Do just northeast of Dong Ha, Marine squads attacked to one side of an enemy machinegun to shoot along its connector trench.

A subsequent base of fire had proven useful in the American Indian wars. Going after automatic weapons from a flank became part of every German Stormtrooper penetration of WWI. Subsequent matrices of each belt got hit from the side in WWII. And, composite guns were outflanked by infiltrators at Iwo and Okinawa. But, not heading right at an offending gun had never been added to the U.S. body of knowledge. Why did participants of that many wars have to reinvent the same tactical wheel? Who or what was responsible for the oversight?

Figure 13.2: Helicopters Brought Casualties Out and Ammo In

(Source: *Courtesy of Mike Leahy, ©*)

14 The War in Iraq

- How did U.S. units maximize their combat power in Iraq?
- What was the unfortunate side effect?

The U.S. Military was well configured for a desert war.

(Source: Public-domain material, from U.S. Army's "War on Terror Images," at this url: http://www.history.army.mil/books/wot_artwork/images/19b.jpg.)

This War Actually Started After the Afghan One

After occupying Afghanistan in late 2001, America didn't invade Iraq until March 2003. But, the Iraqi hostilities would greatly ebb before U.S. forces left the other in August 2021. So, the Iraq War will be covered first. All the while, the Joints Chiefs of Staff had no operational control over any U.S. unit.

> Following the Goldwater–Nichols Act in 1986, the Joint Chiefs of Staff do not have operational command authority, either individually or collectively, as the chain of command

> goes from the President to the Secretary of Defense, and from the Secretary to the regional combatant commanders.[1]
>
> — *Wikipedia Encyclopedia,*
> s.v. Joint Chiefs of Staff

> Today, the Joint Chiefs of Staff have no executive authority to command combatant forces. The issue of executive authority was clearly resolved by the Goldwater-Nichols Act of 1986.[2]
>
> — "About the Joint Chiefs of Staff,"
> www.jcs.mil, August 2022

Safety Had Greatly Impeded U.S. Small-Unit Tactics

Every weapon has an effective range and kill radius. On defense and during the initial part of an offense around 1990, U.S. units started working out elaborate fire plans to bring every weapon to bear on the foe. And most forward-deployed tiny elements were called back in to the friendly formation, so as not to mask its fires. In other words, there were not as many U.S. skirmishers, ambushes, or short-range reconnaissance patrols deployed. This eliminated any need for advanced small-unit tactics. Unlike Eastern counterparts, American squads wouldn't form their own strongpoint or conduct their own assault against a prepared enemy position. Nor would they secretly insert listening posts, create roving sentry posts, perform two-man infiltration of enemy lines, or practice "recon pull" during a non-illuminated night attack. And, this void in advanced tactical technique was badly affecting how America's wars were turning out.

Who Finally Won the Iraq War

U.S. forces showed great daring in their punitive expedition to Iraq. Their troops fought hard at places like Mosul and Fallujah in 2004.[3] But, the whole purpose of war is to restore peace to an entire region, and that takes fully consolidating every rural acre and urban neighborhood. Many of those places would come under additional enemy pressure around 2015.

After Saddam Hussein's army was disbanded in 2003 by Iraq's civilian U.S. administrator, supposedly Sunni guerrillas began fighting for the nation's biggest urban areas. What had been a fairly conventional invasion (with MW overtones), then turned into a counter-insurgency operation (much like the one in Vietnam or Central America).

Yet, there had also been Iranians in Fallujah.[4] And a prestigious Sunni scholar was assassinated in Mosul on 23 November 2004.[5] And, the U.S. Surge of 2007 would confront Muqtada al-Sadr's Shiite Mahdi Army in Baghdad.

So, all the while, Iran had been secretly vying for Iraq. When ISIS (Islamic State of Iraq and Syria) became a problem in 2013, Tehran was allowed to assist the new Iraqi government. The nature of that assistance would greatly affect Iraq's political future

> [U]nits of the *Quds* Force of Iranian Revolutionary Guard Corps (IRGC) were invited by the Iraqi government to help it fight ISIS. By the end of 2017, ISIS' self-proclaimed Caliphate had fallen, but the influence of Iran and its allied groups continued to grow in Iraq. In 2018, a number of those groups formed a political coalition called *Fatah* . . . and achieved a significant victory in the parliamentary elections.[6]
>
> — German Inst. for Internat. and Security Affairs

At some point, Iraq was to become a loosely aligned part of the Iranian Crescent that stretches all the way from Iran through Syria to Lebanon, and now threatens to include Yemen. Tehran had previously dispatched IRGC personnel to help Lebanese *Hezbollah*.[7] It was about to do likewise with Syria's dictator Assad.[8] And the Yemeni Houthis would be next.[9] Those IRGC helpers were good at many things.

A Different Kind of Organization

In 1979, Ayatollah Khomeini had designated—as "guardians of the Revolution"—the *Pasdaran*.[10] (See Figure 14.1.) This unusual outfit became known as *Sepah* or the IRGC. Its ultimate job was to preserve the new Islamic State. "*Sepah* was [at first] a diverse group of guerrillas who [had] initially fought against the Shah

(Pahlavi's) regime, but then joined . . . the successful insurrection of 11-12 February 1979."[11] Its role soon included the following: (1) dislodging other guerrilla factions and restoring order to the cities; (2) suppressing ethnic uprisings in the countryside; (3) providing internal security to expand the Ayatollah's degree of control; and (4) prosecuting the war with Iraq.[12] Embedded in the first four missions was a fifth—gathering "human intelligence."

To neutralize the anti-Khomeini underground, *Sepah* needed intelligence and infiltration. And, to counterbalance Iran's armed forces, it would become an integral part of them.[13] The Shah's "top-down" military subsequently changed into one where every member's allegiance was owed to the Islamic State, instead of his commander. During this Islamization of the armed forces, its ground establishment was converted into a "people's army."[14] Far-Eastern counterparts depend more on light-infantry skills than firepower. So, the Iranian military's focus was about to change.

Sepah was then tasked with training a 20-million-man auxiliary or *Baseej* that was to be Iran's frontline infantry. With the same head count as her "Party of God" (or *Hezbollah),*[15] the two groups may have been one and the same—young men whose Islamic beliefs would come into play on the battlefield. As the war with Iraq progressed, the *Sepah*-led *Baseej* did most of the fighting. Yet, near its end, as many as 80% of the Guard's frontline troops may have been ordinary conscripts.[16]

The Guard's Multiple Mission Was Unique to the Planet

Few of history's other revolutionary armies have concurrently tried to achieve religious doctrine and political stability. How the IRGC chose to perform this mission was the key to its success.

> In rural regions . . . [Iranian] Guard bases [are] located in individual small towns. In more urbanized areas . . . a subordinate headquarters, which may even be a storefront or large house, overseas further subdivisions of the city ("Duties, Aims, Policies of Guard," *Tehran Kayhan,* 14 February 1984, 167). The intention and result is that the Guard achieves maximum penetration of the civilian population, in contrast to the regular military, which only has internal

Figure 14.1: Quds Force Member in Lebanon

Source: Courtesy of Orion Books from *Uniforms of the Elite Forces,* ©1982 by Blandford Press Ltd., Plate 26, No. 77, "Iranian Speial.Forces Trooper 1983.")

security responsibilities in a crisis or emergency (Defense and Foreign Affairs Handbook, 1989, 514).[17]
— *Warriors of Islam,* by Katzman

Mao's Red Guards hadn't operated this way during China's Cultural Revolution of 1966 to 1976. In their search for counter-revolutionaries, they had gotten rid of everything old (customs, culture, habits, and ideas), and then any one with capitalistic leanings. Lacking IRGC structure, those Red Guards had then spun out of control.[18]

But, from deep inside the targeted society, each Sepah detachment was to enforce Islamic law, collect intelligence, quell antigovernment sentiment, and recruit/train soldiers for a "people's army." And, like the Red Guards, it had been delegated enough authority to do all assigned tasks.

> There is . . . no evidence that the Guard has ever had to await orders from the civilian leadership to disperse or arrest demonstrators, but rather the Guard appears to have assumed the authority to suppress demonstrations when they occur. . . .
>
> . . . The Guard . . . has never hesitated to use force against demonstrators, even when those demonstrators were clerics, apparently believing that any opposition to the regime constitutes treason against Islam.[19]
>
> — *Warriors of Islam,* by Katzman

IRGC Control Has Since Been Exported

Sepah's ability to control neighborhoods from within had become widely known to all enemies of the Free World. In 2021, help from Iran for the war in Ukraine was supposedly offered to Russia in the form of missiles. It was more likely *Sepah* squads to help consolidate those areas already occupied. Their method had been already shown to the armies of North Korea and Syria.

North Korea's Light Infantry Training Guidance Bureau (LITGB) personnel can operate against military, political, or economic targets.[20] There would have been a LITGB squad in each occupied South Korean town to counter rebels and develop political cadre. Its effort would have been like that of an IRGC detachment in Iraq, Syria, or Yemen. Assad's slow strangulation of Aleppo almost certainly involved tiny IRGC garrisons. They would have insured the cooperation of all recently captured neighborhoods.

Sadly, this unholy tryst goes both ways. All IRGC personnel have been trained by the North Koreans in UW (of which GW is the primary subset).[21] Most had already been anti-Shah guerrillas.[22] And tiny IRCG detachments had already expanded the Ayatollah's degree of control.[23] That's how those North Korean trainers had learned the powerful IRGC method. It would be the Red Guard version with which the pro-Chinese government of Zimbabwe would get its pro-Soviet rebels under control in 1980.[24]

Then, there was the Syrian civil war involvement. The *Wall Street Journal* of 2 October 2015 reported that the IRGC had some 7,000 IRGC members and Iranian paramilitary volunteers operating in Syria and was planning to expand its presence through local fighters and proxies.[25] In two years, IRGC-affiliated fighters had made quite a difference.

> The Lebanese militant group, labeled a terror organization by the U.S., has grown stronger through its support of the Assad regime, battling Syrian rebels alongside Russian forces and training local Shiite fighters.[26]
>
> — *Wall Street Journal,* 3 April 2017

> *Quds* Force . . . [is an] elite clandestine wing of . . . (IRGC), responsible primarily for its foreign operations. Organized shortly after the Iranian Revolution (1978–79), its activities have centred on organizing, supporting, and at times leading local forces abroad in ways favorable to the interests of the IRGC and Iran's clerical establishment.
>
> The contingent's first activity as a distinct unit came in 1982, when it became involved in the Lebanese Civil War after Israel's invasion of Lebanon. It aided in the organization of *Hezbollah,* a predominantly Shi'i militia founded that same year to repel the Israeli invasion, and remained a close ally and sponsor of the militant group even after the civil war ended in 1990. In the 1990's it turned its attention to Iran's eastern frontier and provided support to the Northern Alliance in Afghanistan against an emergent Taliban.
>
> The unit became increasingly visible on the world stage in the 21st century, after the U.S. invasion of Iraq in 2003 and the regional instability that followed the Arab Spring. In Iraq it played an important role in organizing and aiding efforts among Shi'i militias against U.S. forces, coordinating most notably with the Badr Organization. As the 2011 uprising in Syria devolved into civil war, the *Quds* Force came to the aid of Syria's president, Bashar al-Assad, whose regime was an invaluable ally in Iran's "Axis of Resistance.". . . In Yemen it supported the Houthis, whose rebellion against the central government was bolstered in the aftermath of Yemen's 2011–12 uprising. It also took a leading role in

> organizing ground forces against . . . [ISIS]). Its continued influence in Iraq's internal affairs became a target of popular protests, however, after the *Quds* Force's commander, Qassem Soleimani, intervened in October 2019 to prevent the ouster of Iraq's Prime Minister Adel Abdul Mahdi.[27]
>
> — *Encyclopedia Britannica,* s.v. "*Quds* Force"

IRGC Gen. Qasem Soleimani was killed by a U.S, drone in January 2020, but by then the Iraq government had already been subverted.

How the Iraq War Had Been Fought

Iraq had lain in the Middle East between two hotbeds of fundamentalist Shiism—Iran and Lebanon. Both had a history of self-sacrifice in wartime. Suicide bombers and improvised explosive devices (IEDs) first appeared in 1981 in Lebanon. Such attacks two years later against the U.S. embassy and French and American military barracks in Beirut would force the withdrawal of all Western forces from that country. Those blasts had been arranged by Lebanese *Hezbollah,*[28]

So, when combined with Saddam Hussein's military prowess, this precision way of handling point targets posed a serious threat to any invader.

U.S. forces would practice deception on their road march up to Baghdad in 2003. By avoiding several cities, they had followed a key MW precept. And, they had also applied an unusual form of "recon pull."

Riding specially armed High Mobility Multipurpose Wheeled Vehicles (HMMWVs), the Scout Platoon from a Marine Tank Battalion had ranged far ahead of the maneuver element. Severely outnumbered and outgunned, that tiny contingent did some shooting—but nowhere near what the main force would have produced at the slightest provocation. And with very little collateral damage, the Scouts accomplished a lot from a 4GW standpoint. This drive on Baghdad during Operation Iraqi Freedom I of March 2003 has been chronicled in the TV documentary "Delta Company [1st Tank Battalion]: The Push to Baghdad." But, the personal recollections of the Scout Platoon leader provide more detail on what a small semi-autonomous unit can accomplish.

The [two] Scout sections often worked autonomously . . . but for . . . the drive to Baghdad the whole platoon worked in tandem. . . . The Scout Platoon was comprised of 8 vehicles. The Korean and South African sergeant's sections were mirror imaged . . . [and] comprised of 3 HUMMVs [sic] each. One truck would have a MK-19 grenade launcher mounted with a M-2 .50 in storage. The other truck would have a M-2 .50 caliber mounted with the MK-19 in storage. The third vehicle would have a TOW missile launcher mounted. . . .

The Scouts were the lead reconnaissance unit for the Tank Battalion which was the tip of the spear for the 7th Marines Regimental Combat Team (RCT 7). During this drive northward after Nasiriyah, there were no friendlies to the front of the Scout platoon. The platoon ranged as far as 60 kilometers ahead of RCT 7. In fact, they had already crossed the Nasiriyah Bridge by the time the famous firefight occurred there. While conducting what can best be described as a fighting reconnaissance, it destroyed many abandoned T-55's, T-72's, and assorted other Iraqi military vehicles. Often, it would have to dismount to even approach them. Several other firefights occurred that were mostly one sided. That the Scouts were good at what it did became fairly obvious to the large-unit commanders. That's why one battalion S-3 in Baghdad felt comfortable in issuing the following frag order one morning to the South African Sgt.: "Go forth and do great things and give me a report when you get back." When this platoon got to Baghdad, it was allowed to move freely to cross unit boundaries. After a few days there, the Korean Sgt. was given the task of establishing the Iraqi Police. This was no small task because the day prior their orders were to shoot the police on sight. The South African Sgt. was tasked with gathering human intelligence and acting on it. By this time the platoon was able to capture or destroy many more armored vehicles, fight pockets of Fedayeen Saddam, find and deal with weapons caches, stop several bank robberies, and perform various other [peacekeeping] duties. The platoon use[d] captured Iraqi diesel trucks to fuel generators for hospitals and medical clinics. . . . By the actions of the Marines from Scout Platoon, several of the people of Baghdad they encountered

> would often come to them and offer information the Scouts could act on to make the people's lives safer, earning their trust, and still accomplish the military objectives.
>
> The South African sergeant was able to stay with this unit for five and a half years, whereas the Korean sergeant stayed with this platoon one year longer and is still serving honorably. Their willingness to get up close and personal with the enemy and use [of] compassion when needed had prevented a lot of needless collateral damage.[29]
>
> —U.S. Marine tank scout of South African descent

Overview of the Conflict

In 2003, the U.S. military had been well configured to fight a war across flat and unobstructed terrain, The open deserts of the Middle East provided a perfect locale for this highly mobile and firepower-oriented way of operating. Yet, most units were not properly prepared for the closer terrain of the region—its built-up areas. Many lessons of Manila, Seoul, and Hue City would have to be relearned by motorized and mechanized infantrymen who now found themselves afoot.

Yet, something out of the ordinary did occur during the fight for Baghdad in 2007. An American general asked for outside advice on how to proceed. Posterity Press suggested via e-mail that every neighborhood be segmented into tiny TAORs manned by friendly militias. That way, the opposition couldn't easily shift location, while commerce still flowed in old-fashioned ways. Others probably recommended the same thing, and the U.S. Surge worked. The Shiite Madhi Army was taken to task, and Baghdad pacified when al Sadr agreed to a ceasefire. Unusually fine U.S. leadership had produced the desired battlefield result. Yet, the political future of Iraq was still in doubt

> In October 2010 the whistle-blowing organization WikiLeaks published nearly 400,000 secret U.S. military documents from the Iraq War. . . . They indicated . . . that Iran provided extensive direct military aid to Shi'i militias participating in Iraq's sectarian conflict. . . .
>
> In October [2011], President Obama announced that the remaining 39,000 soldiers would leave the country at the

Figure 14.2: Lebanese Hezbollah Had Been Active in Iraq
(Source: Courtesy of Mike Leahy ©)

end of 2011. The U.S. military formally declared the end of its mission in Iraq in a ceremony in Baghdad on December 15, as the final U.S. troops prepared to withdraw from the country.[30]

— *Encyclopedia Britannica*, s.v. "the Surge"

What Had Finally Happened

Despite innovative warfighting and great bravery by U.S. occupation forces, Iraq had somehow fallen into the Iranian camp politically. A diminutive detachment had led the way into Iraq, but then other small U.S. units hadn't widely enough outposted its cities to avoid their takeover by *Sepah* and its various proxies. The reasons are simple. The Pentagon had thought tiny outposts too dangerous. It was not willing to decentralize its control to that extent. Nor, did it want U.S. infantry squads proficient enough to operate on their own. That would have created more tactical leeway than attack formation participants were allowed.

But, small-unit action is the essence of MW and 3GW in general. And like a *Sepah* neighborhood detachment, those tiny U.S. outposts would have needed highly trained light infantrymen and the authority to conduct 4GW (martial, economic, political, and religious/psychological activity). They could have easily lived off the local economy, helped to create neighborhood cadre, and counteracted unhealthy influences. As Chesty Puller and his sergeant major once did at a small Banana War garrison, they might have even visited the local insurgent commander's office. The most recent example of good outpost design is the CAP program of Vietnam.

But, the Pentagon itself lacks the authority to conduct 4GW. Other U.S. agencies are responsible for the economics and politics of a foreign conflict. Only its martial aspects remain under the purview of the Pentagon, and they are mostly handled in a "high-tech" 2GW manner. Within Iraq, liberal doses of U.S. bombing and shelling were sometimes applied to heavily populated areas, while Muslim extremists focused on community service. So, the Pentagon only waged part of a modern war, while other aspects were left to stagnate.

The Strongpoint Thread Mires in Technology

In Iraq, the average enemy strongpoint had been well hidden in urban terrain. Once located, it was softened up through standoff bombardment before any occupation attempt. Then, most city fighting was conducted across a broad frontage through parallel lanes. This made it hard to outflank a strongpoint. And, there would be only a slight increase in assault element stealth or deception since Hue City.

15 The War in Afghanistan

- How did sophisticated nations lose such a low-tech effort?
- Was the conflict fought in any non-martial arenas?

Afghan battlefield was more precipitous than elsewhere.

(Source: "The Hizara Province," by SFC Elzie Golden, U.S. Army Center for Military History, from this url: http://www.history.army.mil/art/Golden/Image-12.jpg.)

The Irrefutable Conclusion

After America's hurried abandonment of Afghanistan, everyone quickly realized that the U.S. military and its North Atlantic Treaty Organization (NATO) allies had lost another lengthy conflict on the ground.

Those Western countries had specified how the Afghan Army would be trained and subsequently operate. Instead of allowing an Eastern society to develop its own way of fighting from the bottom up, they had insisted on a mirror image of their own "high-tech," firepower-oriented, and overly centralized regimen. As a direct result, the Afghan regime had been unable to control its own countryside. And, too many tiny, semi-autonomous Taliban elements were soon at the gates of Kabul.

Within Those Details May Lie the Outcome of WWIII

Afghan Coalition forces had been predominantly Western—those from America or other NATO countries, and all indigenous ones taught to operate the Western way. Neither the equipment nor tactics of their opposition had been very sophisticated. Its soldiers had known far less about small-unit maneuver than their counterparts in Vietnam, Korea, or WWII. Yet, they still dominated the battlefield. Perhaps, a complementary struggle had occurred in some non-martial arena.

Afghanistan's Assorted Cast of Under-Prepared Players

Afghanistan is a rugged land of Muslim tribal peoples with varying ethnicity. Those who declared a holy war against its *putsch*-installed Communist regime in 1978 became the *mujahideen* or holy warriors.[1] When the atheistic Soviets invaded to bolster that regime in December 1979, many foreigners came to expel them (believing it their Muslim duty). Most were to join a loose alliance of tribes.[2]

> The *mujahideen* who warred against the Soviets were really a motley collection of seven Pakistan-based resistance groups, divided by region, clan, politics, and religious ideology. Worse, the resistance commanders inside Afghanistan had only the loosest of links to the seven groups. For them, party affiliation was merely a matter of access to weaponry—the groups were awash in guns and money, provided by the [Central Intelligence Agency] CIA through Pakistan's Inter-Services Intelligence [ISI].[3]
>
> — *Soldiers of God,* by Kaplan

After facing the well-armed Soviets, many *mujahideen* took occasional refuge in Pakistan's tribal areas. From this remote border region, the Pakistani ISI secretly orchestrated the war against the Soviets. ISI leader—Gen. Akhtar Abdul Rehman—favored the creation of an Islamic State but distributed arms according to each party's strategic contribution, not its degree of fundamentalism.[4] His bureau provided training contact teams, and tactical advisors to the *mujahideen* until 1987.[5]

By 1989, the Soviets had been forced to leave Afghanistan. Their puppet regime stayed on for a while in Kabul, but most of the country was soon engulfed in inter-factional fighting. Then, early in 1994, disillusioned *mujahideen* under Mullah Mohammad Omar and *madrasa* students from refugee camps in Pakistan coalesced to form the Taliban (or "students"). By October, their movement had attracted Pakistani support. Taliban forces soon took control of Herat and Kabul. Then, they offered a home to *al-Qaeda*.[6] Ahmed Shah Massoud—the North Alliance's legendary leader—later claimed the Taliban would use Pakistani advisors as late as mid-1999.[7] Right after 9/11, American forces invaded Afghanistan to drive the Taliban out of Kabul and *al-Qaeda* into hiding.

All the while, the government of Pakistan had officially opposed the Taliban. But, as of January 2007, Islamabad had yet to arrest any high-level Taliban official, and its ISI was still assisting Taliban forces.[8] As a semi-autonomous arm of the Islamabad government, the ISI had occasionally turned a blind eye to drug trafficking to make up for too little federal funding.[9]

This Battlefield Adversary Had Many Weaknesses

The early Afghan guerrilla closely resembled ancestors who had fought the British. He had on-the-job training, a rifle, and a big knife, but seldom followed through with any ground assault. He took the average objective under fire from every side after a half-hearted attempt at infiltration through an unlocked gate.[10] That all-around fire helped him to disguise from which side the main assault might come. Throughout the Soviet-Afghan War, the *mujahideen* contested most little forts this way.

The contemporary Taliban doesn't deviate much from this outpost rejection procedure. Separate elements now alternately cover each other's advance, as in the U.S. Army's "bounding-overwatch".[11]

But, modern-day Afghan fighters still suffer from the same bad habits as *mujahideen*—like too little flank security and ambushing the same places over and over. They also dislike crawling and secrecy in general (due to some masculinity or afterlife issue).[12] Those idiosyncrasies had been known to the part of Pakistan's ISI that guided the *Mujahideen* Council.

> [W]hat was wrong with my method [to the *mujahideen]* was that it lacked noise and excitement. It was not their way to fight, with no firing, no chance of inflicting casualties, no opportunity for personal glory and no booty. Their method was to bombard the posts with heavy weapons by night at long range, move closer to fire mortars, get 30-40 men to surround them, and at short range open up with machineguns, [rocket propelled grenades] RPGs and RLs (rocket launchers). If the garrison withdrew, the posts were captured and the *mujahideen* secured their loot in the form of rations, arms and ammunition, all of which could be used or sold. Then, only then, was the charge laid on the fuel pipeline. If the garrison stuck it out, the pipeline remained untouched.
>
> It often took a serious setback, with quite severe casualties, to force a *[mujahideen]* Commander to review his methods. Like most soldiers the Mujahid hated digging. He was decidedly unhappy in a static defensive role; it was alien to his temperament; it restricted his freedom to move, and he could seldom be convinced of the need to construct overhead cover. Similarly, his fieldcraft was often poor as he was disinclined to crawl, even when close to an enemy position. The hard stony ground, or the possibility of mines, may have had something to do with it, but I had the impression that it was a bit beneath his dignity. Walk, or crouch perhaps, but crawling was seldom acceptable.[13]
>
> — Brigadier Yousaf, Afghan Service Bureau Chief

While the Taliban had some ISI training, most rebels in the anti-American uprising fought like 20 years before. They disliked being fully committed. They shot their weapons just enough to get U.S. forces to retaliate with supporting arms. Then, they stayed well-enough dispersed to keep from getting hurt during their withdrawal. In effect, they disrupted Coalition security just enough to force U.S. troops into a defensive posture.

Before long, Taliban tactics began to mirror those in Iraq. IEDs appeared, and more suicide bombers. There were two possible explanations: (1) more Iranian involvement; or (2) *al-Qaeda* copying *Hezbollah* tactics.

As most of this planet's poppies grow in Afghanistan, the Taliban (and *al-Qaeda*) could afford the best military advisors and equipment available. But, with the exception of anti-air defense, there's little evidence of either being fully exploited. During a Pakistani government cordon in March of 2004, two Chechen fighters were killed trying to exfiltrate an encirclement.[14] A *ninjutsu* trainee would have had no trouble.

The War's Biggest Underlying Aspect

Despite the U.S. military's noteworthy effort, the war in Afghanistan would be largely decided through economic and political factors. When the Americans withdrew into the cities to better protect their personnel, poppy harvesting and opium trafficking flourished in the hinterlands. It produced enough financial gain to influence any number of 4GW issues—to include too little oversight by more than one intelligence agency, and foreign policy concessions by more than one nation.

Drugs had played a greater role in the struggle than most Americans would like to believe. One of the oldest drug cartels in the region—the Haqqani Network—had been behind the most visible Kabul attacks. Haqqani had also been the Taliban's director of operations against U.S. and NATO forces and credited with the introduction of suicide bombers.[15]

When white heroin started to flood America, Washington needed a foolproof defense strategy for the place where 90% of the poppies grew. Only with less smuggling of morphine bricks out and ordnance components into Afghanistan, could that be achieved. Until all traffic along its 14 entrance highways were more closely monitored, no amount of commando raiding or standoff bombardment would get the job done.

During the Soviet-Afghan War, it had been the Pak Army controlling all trucking into Afghanistan, and Pak Intelligence indirectly profiting from any black-market trading. Whoever controlled those trucks at the end of the U.S.—Afghan War may have unwittingly facilitated the drug smuggling. A former American mercenary talks of authorized trucks being (infrared) IR-painted, and the drivers of other trucks sniped.[16] That mercenary had been working for some U.S. government entity.

South Asia is no longer acknowledged by most U.S. agencies as the source of heroin now flooding America. In fact, the 2020 Drug Threat Assessment doesn't even mention Afghanistan. It insists that all heroin comes from poppies grown in South or Central America (most notably Mexico).[17] That heroin may have been fully refined on this side of the Atlantic. But, its opium source was Afghan poppies. It had been likely smuggled across Africa in brick form and then shipped to the Western Hemisphere through Venezuela.[18]

Overview of How the War Was Fought

Afghanistan's hilly interior wouldn't allow as many on-line sweeps as coastal Vietnam. There were some cordon operations, but only to limit some important person's ability to escape. The mountains were full of caverns and below-ground passageways, and the flatland compounds laced with escape tunnels. So, suspected troop concentrations were bombarded, and other places of interest raided by commandos.

But, as in previous U.S. counterinsurgency operations, captured ground wasn't sufficiently outposted to control it. Such outposts were considered too dangerous, and the majority of U.S. forces shifted to the big cities. This had been the same mistake made by the Russians, and any number of post-war analyses admit it. By pulling back into built-up areas, U.S. troops had further sacrificed their technological edge. Urban terrain was too easy to infiltrate, and heat-seeking drones couldn't tell friend from foe.

There were a few U.S. Special Forces squads effectively occupying Afghan villages. The secret to their success had been non-aggressive 4GW-like procedures. According to one Green Beret veteran of both Iran and Afghanistan, not many Yank commandos were expert at light-infantry assault, defense, or E&E.[19] But, they did have 4GW skills. When NATO still sought a village presence, his squad ran a clinic on it. Its soldiers knew how to fight, but instead drank tea with community elders of whatever persuasion. Well enough versed in Asian culture to know "there was a game going on," they carefully studied its parameters. Soon, they could sometimes promote their own agenda. Those village elders had likely been told that NATO wished only a governmental presence, not to kill or arrest anybody.

Such "societal oversight" by students of Afghan custom had been sufficiently straight-forward to succeed. Those special operators were to coexist with local residents for several months without incident. Shortly after being replaced by regular soldiers, one of their patrols was ambushed at the edge of town, thus dooming any further Kabul representation.[20]

But, so diplomatic an outposting had not been totally necessary. A little tactical deception could have accomplished almost as much. *Expeditionary Eagles* shows how a single U.S. infantry squad could—without any artillery shells or drone missiles—rely on ruses alone to defend a roadside outpost against hundreds of attacking narco-insurgents.[21]

U.S. Tactics Through 2005

While Afghanistan's land-locked border was long and precipitous, a network of infiltration routes and waystations already existed at the Pak frontier. When the U.S. first helped the CIA-advised Northern Alliance in late 2001, it badly bombed a formation of Taliban regime and *al-Qaeda* fighters in the Tora Bora region of Nangarhar Province (just north of Paktia). (See Map 15.1.) It then attempted to capture *al-Qaeda's* leader. But, Osama bin Laden escaped the Coalition cordon through tunnels to Pakistan.[22]

Across from Pakistani Wana lay the Afghan province of Paktika (just south of Paktia). From firebases Shkin and Orgun, motorized U.S. troops patrolled the border.[23]

> [W]hat draws U.S. forces to this unexplored part of Paktika . . . is an anomaly of sorts: a sparsely inhabited district with well-tended roads—roads leading to border crossings such as Khan pass that have for centuries served as conduits for Afghan trade, and that today are known to be frequented by al-Qaeda and Taliban fighters.[24]
>
> — *Christian Science Monitor,* 14 January 2004

As had the Russians, those GIs were repeatedly ambushed at chokepoints in the road. The average foe element was in a ravine close to tree cover. After one such encounter, the U.S. reaction platoon narrowly avoided death by paralleling the footprints in a gully. Newly acquired skills of mantracking and deception had served them well. According to 10th Mountain Division headquarters, the opposition had sought GI casualties to down a medivac helicopter.[25] And, that's also why its firebases were being shelled. "The ousted militia . . . frequently fires rockets at U.S. bases . . . , with fighters retreating across the border into Pakistan."[26]

On 8 May 2005, two U.S. Marines were killed while tracking terrorists near Jalalabad.[27] They may have been looking for the exact trace of a major infiltration route. During the Soviet-Afghan War, the best routes originated in Parachinar and then traversed the "Parrot's Beak" (just south of Jalalabad).[28]

Throughout the spring of 2005, U.S. service veterans spoke of considerable danger in Afghanistan. Besides the Taliban, they were now facing the militia of legendary Hekmatyar.

> As a proportion of their total numbers, U.S. troops in Afghanistan recently have been dying at a slightly higher rate than in Iraq. . . . [B.Gen. Greg] Champion said the militants are comprised of more than just the Taliban. They also include smaller elements like the Hezb i-Islami Gulbuddin, or HIG, an extremist group founded by Gulbuddin Hekmatyar, an Afghan who ran training camps in the 1990's.[29]
>
> — Associated Press, 22 May 2005

On 21 May, Taliban soldiers crossed the Pak border to attack a Coalition patrol in Paktika Province.[30] Two days later, Uruzgan Province rebels fired on Coalition forces debarking a helicopter. Others insurgents fled into a nearby cave to escape aerial bombardment.[31]

As the summer of 2005 began, the firefights intensified. Around 1 June, a suicide bomber struck in Kandahar. Six days later, a Pakistani fuel tanker delivering petrol to a U.S. base in Pin Boldak was attacked.[32] On 8 June—in the third incident along the border in a week—militants ambushed a joint American/Afghan patrol in Paktika Province.[33] That's the same day a U.S. base was rocketed near the Pak frontier.[34] On 13 June, a suicide car bomber blew himself up next to a U.S. military vehicle just west of Kandahar, and several roadside bombs were discovered in the city.[35]

There had recently been a precedent of sorts—the attack of a police station in Zabul Province.[36] The tactic was not new to Afghanistan, just to the current insurrection. It was also an essential element of the rebellion raging 1,000 miles to the west. On 18 June, Taliban fighters borrowed another tactic from Iraq—ambushing a police convoy and taking 10 officers and a District police chief hostage.[37]

Four days later, one of the biggest battles since the ouster of the Taliban regime from Kabul occurred at the border of Kandahar

Map 15.1: Afghan Provinces

(Source: Courtesy of General Libraries, University of Texas at Austin, from their website for map designator "iafghanistan_admin-2009.pdf.")

and Zabul Provinces after police discovered a foe camp. Some 50 *jihadists* were killed by supporting arms before they could reach their caves.[38]

On 25 July, rebels attacked a Coalition patrol in southern Uruzgan Province. Another big firefight had occurred in Helmand Province, and an IED attack on an American convoy in Kunar Province.[39]

Countering of the Iraq-Like Maneuvers

There had been car and truck bombs during the Soviet-Afghan War, but none detonated by their drivers. Nor had the original *mujahideen* liked kidnapping. But now, the U.S. occupation was faced with both suicide bombing and hostage taking. The pro-U.S. government had only a token presence in any outlying area. Most in danger were its police and army posts. They were hard to support.

The first suicide bombing had been against Northern Alliance leader Massoud right before 9/11. Then, in January 2004, a man with mortar rounds strapped to his chest self-detonated next to a NATO convoy,[40] and a taxi bomb went off at the funeral.[41] Not until August 2005 were there sophisticated IEDs from Iran,[42] to include (Explosively Formed Penetrators) EFPs.[43] As in Iraq, NATO forces came to expect a vehicular bomb before any urban compound assault after a January 2010 raid on downtown Kabul.[44]

And, like the Iraq war, the Afghan insurgency soon acquired a 4GW dimension. It had taken a heavy toll on the Afghan economy and infrastructure. From the standpoint of politics, local warlords still controlled most of the countryside. And with regard to religion, U.S. troops were still viewed as invading infidels. Had the enemy's strategy changed since the war with the Soviets? Which non-martial influence would be powerful enough to destabilize a Central Asian country without alarming Western capitalists?

U.S. Tactics from 2005 to 2021

In late August 2009, Gen. McChrystal said it would take a new strategy to win in Afghanistan. That strategy would be to protect civilians rather than destroy insurgents.[45] The focus would shift from going after Taliban strongholds to clearing and holding heavily populated areas. Instead of chasing rebels, GIs would create a defensive bastion at the edge of major built-up areas, and then try to understand local, tribal, and social power structures well enough to lessen their degree of corruption.[46] Implicit in the General's plan was less collateral damage.

On 1 December 2009, President Obama decided to send 30,000 more U.S. troops to Afghanistan. His intent was to deny safe haven to *al-Qaeda* and momentarily reverse its capacity to overthrow the Afghan regime. The now reinforced U.S. contingent would become a "security force." Its initial goal was to break the Taliban's hold, and then to secure key population centers. To make sure the Afghan government realized its responsibility to win the war, U.S. troops would start leaving in 2011. In the meantime, they would support all who opposed corruption and hold the door open for negotiations with "good" Taliban.[47]

On 11 January 2010, ABC News carried a story about a U.S. Army company in an outpost just outside a "pro-Taliban" village. This outpost was heavily fortified. Its occupants routinely sallied forth from their fully bunkered movie theater to go on patrol. Then, one day, it was fired upon from the high ground, and a U.S. drone dropped a 500 pound bomb on the suspected source.[48] In effect, the very productive CAP method from Vietnam had not been followed. The village's defense had been moved to its outskirts, but not in as wise a manner.

In February 2020, U.S. representatives and the Taliban agreed to a U.S. withdrawal from Afghanistan by May of 2021. Then, with the Kabul government in eminent danger of collapse, all U.S. forces abruptly withdrew in August 2021.

Afterthoughts

For the first time in history, an American war effort had been defeated in a mostly non-martial manner. There had been no enemy army fighting its way toward Kabul, as had happened at Saigon. The Afghan government had fled the country after its military security simply melted away,[49] One is tempted to credit organized crime for the collapse. Most of the world's opium more easily comes from a loosely controlled narco-state. But, the South Vietnamese regime had also been accused of corruption and poor leadership. Might the mutual trainer of both armies have failed to do something—like properly stress small-unit tactics. For some reason, the Afghan rank and file had lost all faith in their battlefield capabilities.

> Morale across Afghanistan's military ranks was "destroyed" when then-President Trump reached a deal with the Taliban in 2020 and President Biden affirmed the U.S. withdrawal in 2021. . . .
>
> . . . [W]atchdog group has . . . [been] repeatedly warning that Afghanistan's government and military . . . were still reliant on U.S. help.[50]
>
> — *National Public Radio,* 18 May 2022

That last statement could mean Afghan troops hadn't been given enough tactical skill to survive in small numbers. While not part of a fully supported formation, they would have been at the mercy of inveterate fighters. When the Taliban's middle management got set free from detention, the average Afghan soldier felt badly betrayed.

> [T]he Taliban got a huge boost when 5,000 fighters were released from Afghan prisons in 2020. The release was part of the Trump administration's deal with the Taliban.
>
> The prisoner release lowered Afghan soldiers' morale even further. It also . . . raised the Taliban's fighting and organizing capabilities.[51]
>
> — *National Public Radio,* 18 May 2022

Just turned loose were not only well-seasoned combatants, but also small-town cadre. No Coalition soldier would have felt safe where those detainees ended up. The Taliban and their allies were quite brutal religious fanatics. They hadn't been an adversary with whom one could logically negotiate.

> "[M]ost of the released prisoners were group leaders, commanders, and chiefs . . . if sent into a province or a village, they could recruit and mobilize their groups quickly."
>
> [Yet] the U.S. had called the prisoner release a way to build trust.[52]
>
> — *National Public Radio,* 18 May 2022

Any Afghan regular who stayed the course would have risked his whole family. Causing one's foe to lose heart is, after all, the goal of 4GW. So, this final sequence of events may have been previously choreographed.

Some Ways of War Are More Complicated Than Others

Another reason is possible for the Afghan Army's collapse. It takes quite a bit of funding and infrastructure to fight the bombardment-focused American way. By training poor Third World armies to use it, rather than the more economical light-infantry approach, Washington had doomed those armies to failure. Expert light infantrymen need no heavy ordnance to succeed. They rely almost entirely on surprise.

> [O]ther factors behind the failure include: . . . Afghanistan realized that its military had "no logistics and supply capability" just four months before the [U.S.] withdrawal.[53]
> — *National Public Radio,* 18 May 2022

Last Look at the Strongpoint Thread

In Afghanistan, most enemy strongpoints had been fortified compounds. Those identifiable as such were bombarded or raided before any large-scale offensive. But, there would be no increase in the deception and stealth of any American maneuver element, for it now carried small arms with improved target acquisition capabilities. And, enemy machinegun bunkers would continue to be so well hidden, durable, and lethal at short range as to defy any aerial missile, bunker-busting munition, or rifle grenade. Somebody still had to crawl up close enough to find its tiny gunport and then drop in a hand grenade. And, the U.S. would be no closer to how safely to silence its automatic weapon than late in WWII. As at Peleliu, one or two such guns could turn any future landing zone into a death trap.[54]

As before, overall tactical proficiency comes down to bottom-echelon focus and rifleman contribution. Any modern-day American unit that can't sneak up on an alert machinegunner won't easily penetrate a belt of mutually supporting strongpoints. Nor will it eventually defeat any Eastern opponent without excess casualties.

Part Three

Why America No Longer Wins Its Wars

"Success through attrition [America's Traditional Style of War] demands the willingness and ability . . . to withstand attrition, because warfare by attrition is costly [with regard to lives lost]."
– "Warfighting," *FMFM 1*

(Source: "Warfighting." FMFM 1.Washington, D.C.: HQMC, 1989.)

16 Pentagon Has Failed to Evolve Tactically

- What are Americans continually told about their military?
- Is it actually true?

The Pentagon purchases goods and services for U.S. units.

(Source: Free clipart courtesy of pixy.org, image designator "pixy.org/963190/.")

U.S. Forces Still Can't Fully Wage 3GW

MW became Marine Corps doctrine in 1988, and its precepts part of the Army's Air/Land Battle corresponding initiative.[1] But, as the war in Afghanistan drew to a close in 2021, neither service had sufficiently bothered foe centers of gravity (as required by 3GW). Lacking tactically sophisticated point squads, their maneuver elements couldn't extend their avenues of advance deep enough into enemy territory. U.S. special operators had been skilled at raiding soft targets, not assaulting fortified positions. And, all the new targeting devices made long-range bombardment seem adequate.

A deep U.S. ground thrust had been occasionally attempted in Afghanistan, but never behind a 3GW-capable squad. Unless preceded by a little short-range infiltration, no such thrust had much chance of sneaking into, or through, a strongpoint belt.

> Although combat experience should indicate otherwise, the rifle squad currently occupies a relatively minor place in Marine Corps tactical thought. Squad level training and doctrine seem to suggest that the squad has little independent tactical value. The squad has been relegated to the role of subunit whose movements are closely controlled by the platoon commander. Considered in terms of Maneuver Warfare, this attitude is disastrous . . . (Maneuver Warfare demands that the squad assume a primary tactical role).[2]
>
> — Bill Lind
> personal advisor to 29th Marine Commandant

Too Many Lessons of History Had Been Ignored

All major Eastern armies had fully embraced 3GW by the end of WWII. Safely moving through closely arrayed gun nests took tiny teams so highly skilled as to be self-sufficient.

Only out of sheer desperation had American forces briefly experimented with fireteam facilitated 3GW to win some of WWII's final battles. While trying to penetrate a Japanese defense line on Iwo Jima, U.S. forces got routinely pinned down. To limit U.S. casualties, they resorted to two drastic measures: (1) headquarters releasing all control over what happened next to frontline units; and (2) those units allowing pairs of low-ranking volunteers to crawl around behind the foe's automatic weapons.[3] Neither would have been possible under 2GW doctrine.

Then, most forward progress was made through piecemeal effort. At Peleliu, the Umurbrogol's compartmentalization demanded concurrent fireteam maneuving.[4] And, almost every Okinawan hilltop required some short-range infiltration to reach its reverse-slope defense.[5]

Of course, not all Eastern armies have maintained as much 3GW ability since WWII. By 1995, the Soviets had sacrificed much of their small-unit urban agility to technology. After the Chechen disaster, Gen. Mikhail Surkov admitted that "street fighting tactics are absent from the manuals of the Russian armed forces."[6] And, the Chinese had failed miserably with a fully supported U.S.-like punitive sweep into North Vietnam in 1979. Their only adversary had been lightly armed border militia.[7]

The only major armies with sufficient squad proficiency to fully practice 3GW in 2023 are probably those of North Korea and Vietnam.[8]

Nor Can the U.S. Pentagon Conduct 4GW

The U.S. wars in Iraq and Afghanistan were largely fought in non-martial arenas. The Pentagon's exit from Vietnam also had 4GW overtones. Among the VC/NVA ruses in the final assault on Saigon were regular soldiers wearing civilian clothes, special task force personnel dressed like (Army of the Republic of Vietnam) ARVN members,[9] and Communist pilots bombing from South Vietnamese planes.[10] Such things created the impression that a popular uprising was underway. But, nation building from the standpoint of politics, economics, or religion had not been part of the Pentagon's mission. And most non-martial 4GW progress had been through the impromptu efforts of local U.S. occupiers.

Those Pentagon Shortfalls Were No Accident

The West's top-down way of looking at things may produce simplistic strategies, but its upper military echelons still contain very smart people. The Pentagon had put too little emphasis on squad tactics, because it didn't need them. Its steamroller approach to war (with everyone on line and all weapons blazing) didn't require the innovative squad assault methods of WWI Germany. There was no reason for a U.S. formation to surprise anyone. It was more important that composite units mirrored each other, and kept moving forward. Pausing to outflank an automatic weapon or check a hole only decreased that formation's momentum. And any forward-deployed element only masked its fires.

So, U.S. small-unit training de-emphasized stealth and deception. The tiny hand grenade had proven more vital on Okinawa than the largest bomb or shell.[11] But, GIs didn't practice any more grenade throwing after the war. Nor would the Taliban's use of grenades interest the 21st Century trainer.[12] To this day, dummy pineapples for distance are rarely witnessed aboard any American base.[13]

Credit Where Credit Is Due

Fully free U.S. recruits supposedly have more initiative than those from other nations. That's why the American boot camps try so hard to discourage it. More of a group dynamic is their goal. While loud and intimidating, the American war machine is painstakingly slow in its forward progress. That's why one or two GIs so often risk death to get their unit moving. And, instead of a court martial or body bag, they end up with a medal. But, within their desperate measure lies more surprise than the SOPs of the period had made possible. And, throughout American history, there have been any number of ingenious small-unit maneuvers that never made it into the official body of knowledge.

But, such things didn't happen often enough, and therein lies the problem. Other nations—most notably the Germans, Japanese, Russians, Chinese, and other Asian Communists—had already institutionalized enough surprise to accomplish the following: (1) counter Western technology; (2) preserve limited assets; and (3) avoid collateral damage. Since the Revolutionary War, America has enjoyed abundant wartime assets, and been fully willing to expend them. But, as noted this part of the book's introductory quote, one of those assets was young men.

Many U.S. units and sometimes whole organizations have occasionally arrived at better small-unit maneuvers under a forward-looking commander. One was the Marine Raiders of WWII, and another the U.S. Army Rangers. (See Figure 16.1.) The latter are best known for their daring assault on the cliffs above Omaha Beach. But, any unconventional stratagem has gotten little support from their parent bureaucracy; and local tactical excellence been overshadowed by technology, mobility, or some other peripheral issue.

An occasional burst of insight during a long war won't win it. What was discovered must be more widely practiced. But, a Yank CO would say as soon as it's tried, it can't work again. Meanwhile, his Eastern counterpart reasons that if enough deception were added, a firepower-loving opponent would never spot the core procedure.

Figure 16.1: U.S. Rangers at Omaha Beach
(Source: *FM 100-5* (1994), p. 52)

That's why enemy sabotage in Vietnam so often looked like a lucky mortar hit. Might thousands of lone sappers have so often fooled the American commanders that Congress got tired of funding the destroyed ordnance?

The Moral High Ground

The Pentagon has always striven to do the right thing. It takes the straight-forward, steamroller approach to war to end the bloodshed as quickly as possible. In fact, all Western militaries have experienced fewer integrity violations than Eastern counterparts since the days of Sun Tzu. That's why modern-day Asian Communist armies try so hard to remove any exaggeration from their after-action reports. But, that direct a U.S. onslaught also creates a lot of collateral damage and too big a "human target" for long-range automatic-weapons fire. Even when fully mechanized, GIs pay a terrible price for their boldness.

To further complicate matters, not all wartime ruses are morally reprehensible. According Pope John Paul II, battlefield morality is as closely associated with how many casualties are inflicted, as are suffered.[14] Every time a war is lost, all friendly losses will be in vain. Yet, the U.S. military still insists it has more regard for its soldiers' lives than any Eastern army. Americans units have seldom "given ground," while Eastern foes regularly fall back to save lives. U.S. soldiers do as told, while Japanese or German soldiers are permitted more initiative. And, GIs make no personal contribution to their manuals, while Red Chinese troops monitor parent unit histories. So, which side more often holds the moral high ground is far from clear.

17 Shock and Awe Versus Deceptive Evasion

- Are machines more productive than people in war?
- Which do the Asian armies like better?

Big enemy guns can be hidden, then remotely aimed and fired.

(Source: Podgotovka Razvegchika: Sistema Spetsnaza GRU, © 1998 by A.E.Taras and F.D. Zaruz, pp. 374, 375.)

Washington's Next Mistake in Judgment

More problems surfaced when the governmental agency traditionally supplying the U.S. military took over all battlefield responsibilities—to include maneuver. As the Free World's arsenal, it had helped to win WWII. But, its big-picture thinkers then concluded that—by expending more ordnance herself—America could win a personal dispute. In some cases it did, and in others not. Additional infrastructure got damaged, complicating the peace process. Additional civilians and GIs got hurt by mistake. And, additional enemy soldiers dispersed without ever emerging above ground.

Precision munitions do no good where the enemy's precise location cannot be determined. And even then, the type of ordnance matters. On Okinawa, it took a dozen armor-piercing 155mm rounds in the same spot to disable a concrete pillbox.[1] U.S. grenades were effective, but only when someone could sneak up close enough to a hidden aperture to drop one in. And, the details of when to stress maneuver over attrition didn't survive the truce. Nor would the arms manufacturers' interest in "big-ticket" items.

Overseas Conflicts Aren't Decided Through Bullets Alone

Firepower and surprise are interchangeable in war.[2] Every time defenders are amazed, less shooting will beat them. Where fewer weapons suffice, so can fewer attackers. That means not as many friendly lives at stake. Plus, only a tiny maneuver element can get that close anyway. In effect, better tactics generate more surprise. Just looking good won't get the job done without excess casualties. Most safely tackling an enemy objective takes stealth and deception. Too much equipment makes noise in transit.

Asians More Readily Admit to Gear Limitations

When the location of a fortified target is in doubt, no amount of standoff bombardment will destroy it. The big Oriental nations have other ways of solving that problem. Historically, it was often done through infiltrating snipers doubling as forward observers.

> It is people, not things, that are decisive in war.[3]
> — Mao Tse-tung

The excuse normally proffered by Washington is that trans-Pacific nations can't afford fancy equipment. They may lack the funding to develop it, but are quite good at stealing or re-engineering it. Though not as wealthy as Western counterparts, they still produce excellent airplanes, ordnance, and rolling stock. And, their primary focus remains the soldier's mind and spirit.

> Sun Zi [Sun Tzu] . . . stressed human beings' dynamic initiative in changing the disadvantaged [situation] into the

> advantaged, and the passive into the active.[4]
> — *The Strategic Advantage*
> New World Press, Beijing

> In . . . regular warfare, . . . it [armed Communist struggle] is fully capable of . . . getting the better of a modern ["high-tech"] army. . . . This is [through] the development of the . . . military art, the main content of which is to rely chiefly on [the] man, on his patriotism and . . . [individual] spirit.[5]
> — Gen. Vo Nguyen Giap

But, it wasn't just Oriental societies sharing this focus. Germany and Russia would both become interested in it. By the mid-20th Century, Japan and all Asian Communist nations had their own version of Stormtrooper technique and more emphasis on *mano-a-mano* combat than Eastern European advocates.

> Japanese strength...lay in small units and the epitome of Japanese doctrine was embodied in small-unit tactics. Night attacks...and the willingness to engage in hand-to-hand combat were the hallmarks of the Japanese infantryman. Indeed such tactics were very successful against Soviet infantry [in 1939].[6]
> — U.S. Army, *Leavenworth Papers No. 2*

Japan's small-unit aggressiveness would fail to win the war in the Pacific, but not because of U.S. equipment. The broken terrain of those islands had prevented tanks, bazookas, and heavy artillery from bothering many defenders. Then, in Vietnam, the NVA needed only a few amphibious tanks or long-range guns, and no aircraft to hold their own against the U.S. phalanx. Once again, the reason was people. Thousands of near-naked sappers had blown up enough ammunition and fuel to limit America's use of supporting arms. They did so by emplacing timed explosives during a preliminary mortar attack.

Proof of Such a Claim

The USMC An Hoa ammunition dump south of Da Nang had gone up like a miniature atomic bomb shortly after dawn on 23 February 1969. It's loss was soon attributed to indirect enemy fire.

Yet, one independent researcher claims enemy troops had entered the wire and lobbed satchel charges into the dump with a bamboo catapult.[7] While satchel charges were likely involved, such a risky delivery method would have lacked the precision for a fully bunkered target.

According to the 5th Marines' *Command Chronology*, "Incoming rounds had hit around the ammo dump at 0140, but the dump did not blow until 0610." The *Command Chronology* of 3rd 155mm Gun Battery shows enemy rocket and mortar attacks in the vicinity of the dump at 0130, 0700, and 0730.[8]

So, the final An Hoa explosion must have been the work of a VC sapper. Only he could have reached the buried munitions. The residents of An Hoa's nearby village are known to have secretly tunneled right up the base's barbed wire.[9]

Then, at Biên Hòa Air Base in 1972, after witnessing several "overly effective" mortar attacks against American aircraft over the years, U.S. Air Force Police finally connected the dots. It was supposedly the bomb storage area under attack, yet no indirect-fire impacts were reported at that location.

> A sapper attack took place on 12 Jan. 1972, resulting in the explosion of the ammo dump at Biên Hòa Air Base. On that fateful night, explosions began at the bomb dump. . . . We initially thought it was another rocket attack but no one reported impacts or explosions except in the ammo dump.[10]
>
> — USAF 3rd Security Police Squadron report

Backdrop to Such Events

The total number of parked planes and ammunition dumps lost throughout the Vietnam conflict is hard to determine. Only known is that indirect fire damaged "too many" aircraft at Biên Hòa Air Base in 1964 and Da Nang Air Base in 1967. Then, it destroyed both the ammo dump and too many aircraft at Camp Evans in 1968. All the while, sappers are known to have gone after aircraft at Camp Holloway Airfield in 1962, Da Nang Air Base in 1965, Marble Mountain Air Facility in 1965, and Kontum Airfield in 1968. Enemy mortars or rockets landed while such sappers attacked at Chu Lai Air Base and Marble Mountain Air Facility in 1965, Chu Lai Air Base again in 1968, and Cu Chi Army Airfield (near the tunnels) in 1969.[11]

Still, the fully refined enemy ruse took a while to develop. At Chu Lai in 1965, the planes had been sprayed with "Tommy Gun" fire as satchel charges were tossed up their tailpipes. The use of a timer didn't become likely until a "seven-helicopter-ruining string" of mortar rounds at Marble Mountain,[12] and the An Hoa blast as a few tardy mortar shells landed.

American artillery pieces and shells were also a favorite target of the NVA. Satchel-charge-throwing sappers did much of the damage at Firebases Maury I in May 1968, Airborne in May 1969, Henderson in May 1970, and Mary Ann in March 1971. At Henderson and Mary Ann, foe mortar rounds preceded the sappers.[13]

In all of the above cases, enemy personnel had somehow snuck between GI sentries. Others had conducted Stormtrooper-like assaults.

U.S. Response

Most U.S. base commanders never knew they were being hoodwinked. Nor did they try to catch the occasionally visible intruder.[14] Instead, they worried about being overrun. Apparently, they had never heard about Chinese Communists "using 'ordinary forces' to engage and 'extraordinary forces [infiltrators]' to defeat."[15]

> When comparing the two sides, their strength and weakness should be considered in specific conditions, chiefly the actual effectiveness of strength in action and in the confrontation. A strength in action should be considered in . . . human dynamism (resolution, capacity, resourcefulness, creativeness) which are very decisive factors.[16]
>
> — National Political Publishing House, Hanoi

The Ongoing Competition Between Mind and Materiel

To offset the Taliban fighter's inadequacies, the suicide bomber was introduced to Afghanistan. Objectives that couldn't be infiltrated were opened through a driver-detonated truck bomb. Of course, suicidal fighting was far from new. On Okinawa during WWII, only slightly fewer Americans were killed offshore by *kamikazes* than through ground action (4,907 vs. 7374).[17] There were rumors of suicide bombers in Vietnam, as evidenced by the movie "Platoon."[18] Then, in the Middle East, *Hezbollah* perfected the method around 1979.[19]

The suicide bomber was to be the poor man's drone missile near the end of the 20th Century,[20] just as the sapper had been his "eyes on target" during the Vietnam War. While mostly for political effect,[21] the self-destructive human being also became a way for Muslims to fulfill their fatalistic beliefs. When combined with lucrative drug running, he led to a well-rounded 4GW initiative.

But near-suicidal effort has never been officially encouraged within the U.S. military, nor should it be. For every U.S. soldier or Marine, the constitutional right to life, liberty, and the pursuit of happiness must include every opportunity to come home alive from a foreign war. In this age of more lethal enemy weaponry, that takes smaller maneuver elements. They're harder to detect and more difficult to destroy.

18 Some New Hardware Discourages Maneuver

- What are GIs continually told about their new equipment?
- Does it all make it easier to win at war?

Modern electronics will also remove the screen of darkness.

(Source: FM 7-8 [1984], p. 3-57 .)

Combat Readiness

All U.S. Marines must maintain a positive attitude. It was their trademark in WWI, and has since produced much of their finest heritage. Most U.S. Marine Corps (USMC) infantry outfits work on any "glitch" in house so as not to bother headquarters. But, here's the catch. The most common challenge has been around since 1918—a lack of maneuverability at the squad level. Nor is HQMC apparently aware that such a deficiency prevents the full practice of its new doctrine—MW. Only with a highly skilled spearpoint can its maneuver elements flow around enemy strongpoints or safely penetrate a few. All the while, its trusting flock believed all lessons learned and new weaponry to have been thoroughly considered. They weren't.

In fact, top-down military bureaucracies eat lessons learned. Any tactical improvement conflicting with standard procedure gets easily lost in the shuffle. And if that organization occurs in a capitalistic society, most infantry outfits will be continually barraged with the latest equipment. How to use all those new gadgets then takes precedence over maneuver practice.

Not All Changes Will Lead to Progress

Most of the refinements to America's wartime arsenal are for fatally injuring more opposition soldiers. They further force the enemy into secretly shifting location. On defense, he digs deeper, and on offense he approaches his objective below ground. Either way, the U.S. unit tries to overpower, rather than outmaneuver, him.

But, that new weaponry can also act as a crutch. What the American soldier did in its absence no longer matters. Each WWII Raider had his own feints for hand-to-hand combat.[1] Today's riflemen no longer count on any foe getting that close. Who is safer? And, the latest ordnance may not be fully tested. Nor is it immune to damage, moisture, or dirt. So, backup moves might come in handy.

Other hardware more directly affects maneuver. Combat units used to move along linear landscape features like streets of a city. Many had travel or blocking characteristics. Exploiting them was called land navigation by "terrain association."[2] Now, such units go straight to the next objective or follow arbitrary GPS legs. Through "dead reckoning," they take little advantage of intervening real estate. Not surprisingly, most transit legs are blocked by easily defended gullies and ridgelines.

Machinegun Was Last Ordnance to Greatly Alter Warfare

The Allied tank/infantry sweeps of WWI were fully exposed to automatic-weapons fire from the front. Tripod-mounted German

Maxim MG 08's could accurately dispense 500 rounds per minute, enough to scythe down successive waves of Doughboys like wheat in a field.[3]

In the Pacific during WWII, the Americans took machinegun fire from all sides as they tried to penetrate a belt of squad strongpoints. Such guns could only be approached by crawling to one side, and then assaulted from the flank or rear. Anything less was "pre-machinegun" technique. And, all squad assaults should have been amended worldwide. But, only in the Eastern World did that happen. Elsewhere, those in charge thought it wiser to suppress all known machineguns and then rush in with their own weapons blazing—namely, upright and on line. But, so doing didn't work very well because a few hidden weapons survived every barrage. And it only took one such gun to knock down any number of upright attackers from a distance. At Stalingrad, a pair of Russian machineguns at a crossroads house kept an entire German battalion from advancing.[4]

On Peleliu, the Japanese created sheafs of plunging automatic-weapons fire to supplement their artillery concentrations. A .50 caliber machinegun bullet can travel up to 4.5 miles. Halfway across the huge airstrip clearing, on-line elements of the 1st Marine Division could only move a little faster to escape the "sideways stream of bullets." The hard coral surface permitted no other option.[5]

Night Vision Devices Have Made Little Difference

Some animals see well at night. But, most humans must adjust to the dark for up to 40 minutes before their limited scoptic vision takes over. Those who are color blind better distinguish shades of grey. And, there are tricks for seeing more at night, like moving one's head from side to side to further examine some object through their peripheral vision. *Ninjas* have eye exercises to increase that sideways capability,[6] and ways to examine the darkness through unexplained shadows *(ankokutoshijutsu)*.[7] The *ninja* is also alert to all ways of being detected. He formerly stayed in shadow, to the front of some dark object, or outside the sentry's field of vision. Now, he waits for some light to blind any night vision device. From the direction of a low-slung moon, he approaches a defense. And to pass any sentry, he waits for a cigarette to be lit.

Nor can the latest thermal imaging see through heavy vegetation or ground swells. To close with a Western objective, the Eastern soldier has only to move through the rainwater erosion. And to cross an open expanse, he keeps a distant flare at his back.

To realize the limitations of night vision equipment, one must know some history. Under-gunned Eastern forces more often fought at night. During WWII, the Japanese thought U.S. soldiers night-blind. Their own Privates had done eye exercises since 1913.[8] And being able to see is only part of the skill. Also important is how to move. Secretly traveling at night takes staying off the horizon, because the smallest mound can provide a natural horizon to any defender. And light-colored surfaces give him telltale backdrops. To avoid reflecting light,[9] *ninjas* shift between shadows. The interaction of light and shadow is called *chiaroscuro.* The brighter the light, the darker seems the shade. The shadow warrior takes can predict where the most shade is likely to occur. He can approximate its length.[10] And, he knows that a sentry is momentarily blinded by a cloud drifting over the moon. Just before sunset or after sunrise is his best time to move. That's when his adversary's eyes have trouble adjusting between bright sky and dark ground.[11] To avoid being seen from any direction, the careful traveler keeps poorly illuminated objects to his opposite side.[12]

Both Seeing and Moving Are at Issue

Optoelectronics have been around a long time without much affecting any battle. U.S. forces had great difficulty approaching the Shuri Line on Okinawa in WWII. Yet, each U.S. Army division had 110 sniper scopes and 140 snooper scopes, both of which could see by means of infrared radiation. According to an official battle chronicle, "the former were mounted on carbines and permitted accurate night firing, while the latter were on hand-held mounts and could be used for night observation and signaling."[13] Throughout the losing effort in Vietnam, American grunts had starlight scopes, and U.S. tankers prototype thermal imaging.[14]

The problem was simple. All big Eastern armies (particularly the Asian Communists) have constantly searched for ways to outsmart Western technology. While lifting the siege of Thuong Duc in the autumn of 1968, 2nd Battalion, 5th Marines was deployed as a diversion at the mouth of a big mountain valley south of Da Nang.

Figure 18.1: NVGs Hinder as Much as Help
(Source: FM 7-8 [1984], p. G-4.)

Shortly after seizing a pyramid-shaped peak at the foot of Charlie Ridge, Golf company got painfully introduced to world class night fighting. In full moonlight on a sparsely vegetated hillside, NVA soldiers dressed like bushes got within grenade range of American sentries by moving slower than was perceptible to the human eye.[15] That much surprise in 1968 should serve as a warning of how future foes may continue to circumvent the latest electronics.

So, who can be expected to do better after dark—the under-equipped enemy soldier familiar with fighting at night, or the U.S. soldier trying to dispel the darkness? Certainly, the night vision goggle (NVG) better helps defender than attacker. But in heavily foliated terrain, it's of little use to a perimeter guard. If fire lanes have been cut, he may occasionally see an upright assailant. But, one crawling through watershed ditches would go undetected. At Guadalcanal, all Japanese patrolling near U.S. lines was done in the prone position.[16]

On offense, thermal imaging may help to locate listening posts during a preliminary reconnaissance. But, during the assault itself, infrared radar could compromise surprise, or too many facial gadgets block one's vision. (See Figure 18.1.)

Thus, it would be incorrect to say that the latest night vision gear makes U.S. troops competent after the sun goes down. Such a conclusion only enhances somebody's profit margin. It takes seeing, hearing, smelling, touching, tasting, and their subliminal combination to do well after dark. Most Eastern combatants attempt to remain hidden, and keep one eye shut during the slightest illumination. In that way, they take full advantage of their God-given abilities. Only when U.S. troops get to choose which gear to wear, and have recently practiced obscure movement, will they do well during night operations.

Tank Support No Longer Vital as Well

As evident in Ukraine, the main battle tank must now be more sparingly used. Improved guidance systems have made the shoulder-fired munitions too lethal. But, the Javelin missile has a minimum arming range of 65-75 meters.[17] And in broken terrain, tanks are most vulnerable at a shorter distance. During the latter stages of WWII, the Chinese, Japanese, German, and Russian infantries all practiced barehanded tank killing. The Asian Communists later adopted several variants. Because much of the threat came from below ground, the armored vehicle's best defense was a close-at-hand Yank rifleman on all four sides. In Vietnam, U.S. tanks blew a track before getting anywhere near a big firefight.[18] Most likely, a VC spiderhole occupant had used communication wire to drag a tank mine under one of their treads.[19] Both the Germans and Japanese did something similar in WWII.[20]

The Paradox of Precision Fire

HQMC now bases much of its field training on computer war-gaming with precision munitions. Virtual combat has many inherent risks. With regard to 4GW, the term "precision fire" is largely a misnomer. The only way to preclude civilian casualties is a U.S.

ground presence nearby. From 20,000 feet overhead, a dress can look like a robe. And from 1000 feet along the ground, a book bag can seem a satchel charge.

America's preferred way of war—"high-tech 2GW"—is also how her top-level military leaders view the battlefield. With standoff weaponry, they go after concentrations of troops, guns, and supplies. But, what if the enemy has fully dispersed all assets. Then, the U.S. can no longer muster enough planes or artillery in any one spot to engage all targets. That happened all the time in Vietnam.[21] Or, if the opposition is within danger closer range of friendlies, he is protected by the safety factor.

In Chinese Mobile Warfare, tiny assault elements separately approach tan objective to later reassemble.[22] Enough fake artillery makes silencing the real ones nearly impossible (as in Ukraine).[23] And supplies can be issued to the most needy unit from a sub-surface depot (like in 1953 along the 38th parallel [24]).

So, continually monitoring a contested area and then killing whatever moves is an unrealistic way of controlling it. The Cu Chi tunnels permitted any number of enemy fighters to approach Saigon from Cambodia. And, no future surveillance system will likely peer below ground from an airborne platform.

Now It's the Killer Drones Getting the Headlines

Of all the wartime introductions to Ukraine, the drone has shown most promise. But, drones can carry an unwanted message. Too much aerial surveillance in the same spot means an impending incursion. After losing a patrol that way in 2000, Israel temporarily suspended all commando raids to Southern Lebanon.[25]

The U.S. has reconnoitered by drone since early in the Vietnam War. But, so doing had little—if any—effect on its outcome.[26] Against a well-camouflaged foe, unmanned aircraft see little more from immediately overhead than was previously possible from higher altitudes. Just as "precision fire" from aircraft and artillery creates too much collateral damage, so will fire from an unmanned drone.

What Had Happened to Early America's Tactical Insight

The excellence of an infantry maneuver can be best measured

through casualty avoidance. No friendly losses whatsoever takes totally surprising the foe. A large unit would need highly skilled small units. On offense, it must have buddy teams that can sneak between sentries unnoticed. On defense, it must have squads that can form self-sufficient strongpoints.

"Advanced" squad stratagems are what the Japanese, German, Russian, and Chinese armies collectively developed in WWII—by all trying to counter superior firepower. By 1945, their small-unit maneuvers had become almost identical for most combat scenarios. Because not all four nations were allied, that same way of fighting became (and still is) the state of the art for squad tactics. This paradox of history has been confirmed by a former CO of the U.S. Marine Corps' East Coast School of Infantry.[27]

In only a few American units had "ideal moves" been attempted out of sheer desperation. That's because most Yank outfits thought their own firepower more important. And at some point after WWII, those safer ways of operating were replaced by technology in the Chinese and Soviet armies. Only at two places do they still exist—Vietnam and North Korea.

The Most Obvious Sacrifice

As in martial arts, every battlefield participant needs a certain amount of agility to escape common dangers. He must drop down to avoid shrapnel, and get behind something to escape bullets. Yet, American infantrymen have been regularly draped with as much equipment and ammunition as they can carry. (See Figure 18.2.) According to S.L.A. Marshall, if a soldier's load in the approach march is more than one third his body weight, he must rest before attacking.[28]

It is in this approach march that the extra weight/bulk first becomes problematic. It makes the wearer less stealthy and alert. Then, in the assault, it hinders his fire and movement. During a chance encounter in Vietnam, young Americans quickly jettisoned all packs and extra ammunition. This left them with a web harness, flak jacket, weapon, and helmet. Anything protruding from that harness made crawling difficult. The flak jacket helped against shrapnel, but not rifle bullets. Nor will the modern-day Kevlar armor stop bullets without a special ceramic plate.[29] For a while in Vietnam, deployed members of 3/5 wore no flak jackets at all to increase their battlefield effectiveness. The M-16 rifle was lighter than the M-13, and its ammunition as well. But, two full years after the M-16 was introduced, it was still jamming after the first shot on full automatic.[30]

America's SWAT policemen can assault behind a bulletproof shield, but her grunts enjoy no such luxury. To avoid getting shot, they have to outshoot the defender or dodge his bullets. Any head harness or extruding gear makes that difficult. Even restrictive clothing can be counterproductive.

GIs Are Not Hardened Weapon's Platforms

U.S. war planners must further decide how important the ground combatant's field of vision is. Ever since WWII, the opposition shooter has been hard to spot. Any NVG obstructs not only the wearer's peripheral vision, but also his far and near frontal perspective. Where the opposition can now emerge from any range or direction, such a limitation could prove lethal. Too much radio traffic also makes it hard for the rifleman to detect telltale sounds—like an AK-47 coming off safe. That's why hand and arm signals used to work better than verbal instructions. In effect, any enhancement of one sense may cause others to be misinterpreted.

The Yank grunt is still expected to slowly advance against long-range machinegun and artillery fire. Of late, he rides around too much to get enough practice at how to safely operate dismounted. And now, he has thermal imaging and subterranean dwellers to worry about. The best way to counter such threats is to crawl through the microterrain. But, his new gear provides no protection from getting spotted. Most makes the prerequisite crawling and alertness more difficult. In effect, the technologically advanced U.S. rifleman can less evade harm than his WWII predecessor. Too much new gear has prevented him from becoming more tactically proficient. That may be why "boots on the ground" are no longer popular in Washington. But, as in Ukraine, the inside of an armored troop carrier can prove even more deadly.

The Modern-Day Elephant in the Room

The latest fad in America is the cell phone. To feel safe, pretty women maintain a distant contact while walking through dangerous neighborhoods. Cameras may record what happens next, but help won't get there in time. As with the lowly grunt in enemy territory, closer attention to one's immediate surroundings would have done those women more good.

For the first time in history, ordinary grunts may be able to talk with their fireteam or squad leader by radio or text message. (See the tiny mouth piece and eye screen on Figure 18.2.) At that point, a more complete picture of his unit situation becomes available to all leaders. But, if too much time is spent on how to handle that situation, it will change; and the opportunity for momentum be lost. So, most "master minding" must occur beforehand, and a somewhat appropriate technique applied. Then, if every squad member can

Figure 18.2: Land Warrior
(Source: Courtesy of Raytheon, © 1999)

slightly alter his role (as new obstacles appear), that method will work. But, no battlefield dilemma can be solved quickly enough through cell phone traffic on the spot.

Within Asia, every gap may be a trap. Any U.S. squad that hasn't tactically prepared for that eventuality will be at severe risk. No modern-day gadget will save the day.

19 Squad Tactics Still Vital to Modern War

- Why are tiny units so necessary in "high-tech" combat?
- Which of their maneuvers work best?

Modern targeting equipment puts large formations at risk.

(Source: Courtesy of Sorman Information and Media, from "Soldf: Soldaten i falt," ©2001 by Forsvarsmakten and Wolfgang Bartsch, Stockholm, p. 225.)

The Inescapable Trend in Contemporary Warfare

With the onset of the killer drone, every military unit is in imminent danger from the sky. Its only defense is to shoot down the drone, hide below ground, or disperse into elements too small to make lucrative targets. The self-sufficiency of such an element will depend upon its ground signature and combat power. The smallest group that can forcefully withstand a human wave or penetrate an enemy cordon is the infantry squad. But, easiest to obscure is the individual rifleman. And he can quietly sneak through an encirclement.

How the Value of Human Life Comes into Play

In an Asian Communist nation, the individual soldier's life has less value than in a Western democracy. One society is atheistic, and the other Judao-Christian. So, the PRC, North Korea, and North Vietnam are often accused of too readily expending their soldiers. But, what looked like a Chinese human-wave at the Chosin Reservoir was really extended squad columns on line, each trying to make its own penetration.[1] And North Korea has provided ordinary riflemen with *ninjutsu* skills. Its "Light Infantry Training Guidance Bureau" likely detaches—as training cadre to each line company—a few accomplished commandos.[2] Throughout Korean history, the *hwarang* have been used as scouts, guerrillas, shock troops, and night infiltrators.[3]

But, WWII Japan was a Shintoist military dictatorship. Confucianism had helped to shape Shinto through its emphasis on responsibility to the community and obedience to elders.[4] That may be why Japan's Shinto would later stress "family respect and the subordination of the individual before the group."[5] Its emperor made a perfect father figure. To his people, he was almost like a god. So, during WWII, giving one's life for that emperor seemed perfectly natural. And, many Japanese soldiers would be lost to *banzai* charges or some other suicidal effort. Their code of *bushido* had been so rigid that U.S. propaganda soon characterized them as rabid beasts in need of extermination. Such demonization facilitated America's traditional style of war—killing as many enemy combatants as possible.

First Impressions Not Always Correct

Before too quickly criticizing a foe's manpower thrift, one must make sure his own soldiers are being given every opportunity to survive. Within both Shintoism and Confucianism, human beings were considered inherently good. Neither belief system admits to original sin or that all humans are initially flawed (like Christianity).[6] Yet, as most total-war participants, Asian Communist and Japanese infantrymen had few plans for the future. And similar fatalism would soon be observed in the radial followers of Abrahamic Islam.

Such talk makes the American Judao-Christian infantryman the best motivated in the world, but not necessarily the most skilled. While highly respectful of his own life, he is also tempted—through his traditional way of war —to under-regard the life of opponents. And so doing places his own moral integrity in peril.

Japanese, Communist, and Muslim leaders may not have valued each soldier's life as much as American leaders did, but they did assign more strategic significance to that life. As a result, their infantrymen were better trained tactically than Americans. Japan had been instructing its riflemen on sensory-enhancement, shadow-walking, and other *"ninja"* techniques since 1913.[7] And, there had been *jihadist* training camps throughout Afghanistan, Pakistan, Iraq, Iran, Syria, Lebanon, and Sudan.[8] Videotapes of *al-Qaeda* camps show individual fighters tackling live-fire obstacle courses. They had been building the hand-eye coordination and muscle memory to do well against a better-armed opponent.[9]

Meanwhile, in a top-down and highly prideful Western outfit, the recently enlisted and quickly trained rifleman can seldom take cover from incoming fire, much less hide. Nor is he shown how to sneak between enemy holes. As suggested by Chapters 9-15, he more often moves upright through long-range machinegun fire.[10] So, which side has been more thoroughly preparing their young men for battle is a matter of opinion.

Advanced UW Has Two Basic Strategies

The UW portion of 3GW heavily depends upon infiltration on offense, and ambush on defense.[11] In the last big battle of WWII in the Pacific, no Japanese occupied hilltop could be seized without some infiltration of its reverse slope defenses.[12] And, whenever the outgunned NVA/VC chose to fight in Vietnam, it seemed to the Americans like an ambush.[13]

Short-Range Infiltration Is Not That Hard

Because U.S. forces have usually attacked *en masse* during the day, most GIs think an American sapper attack too risky. That's not the case. Where the quarry's fields of fire have been partially cleared or sentries put on reduced alert, U.S. sappers could sneak into an enemy objective at noon. If Eastern Hemisphere troops can do something, so can those from the Western Hemisphere. The former have no more patience. To develop real patience, they would have to spend time at the bottom of a Western bureaucracy. Below is a rare account of how "natural" such an attack feels to any participant.

> Once [in the Mekong Delta in 1965] I got into the middle of Cai Be [ARVN] post where the district chief's office was. It was fifteen days prior to the attack and takeover of the post by our battalion. I was accompanied by two comrades armed with submachine guns to protect me in case my presence was discovered while I was nearing the post entrance. I was then wearing pants only and had in my belt a pair of pincers, a knife, and a grenade. At one hundred meters from the post I started crawling and quietly approached the post entrance with the two comrades following me. At twenty meters from the post, my comrades halted while I crawled on. At the post entrance there was a barbed-wire barricade on which hung two grenades. Behind the barricade stood a guard. I made my way between the barricade and the stakes holding up the barbed wire fence. I waited in the dark for the moment when the guard lit his cigarette. I passed two meters away from him and sneaked through the entrance. On that occasion I was unable to find out where the munitions depot was but I did discover the positions of two machine guns and the radio room. I got out at the back of the post by cutting my way through the barbed wire.[14]
>
> — Reconnaissance sapper
> 261st VC Battalion

Short-range infiltration is safer than other forms of infantry assault. That's quite difficult for desk- or vehicle-bound leaders to realize. So, Figures 19.1 through 19.16 will hopefully reveal what a pair of GIs might experience during a squad-sized sapper-like attack.

Figure 19.1: Attack Objective from a Distance
(Source: FM 7-8 [1984], p. 69.)

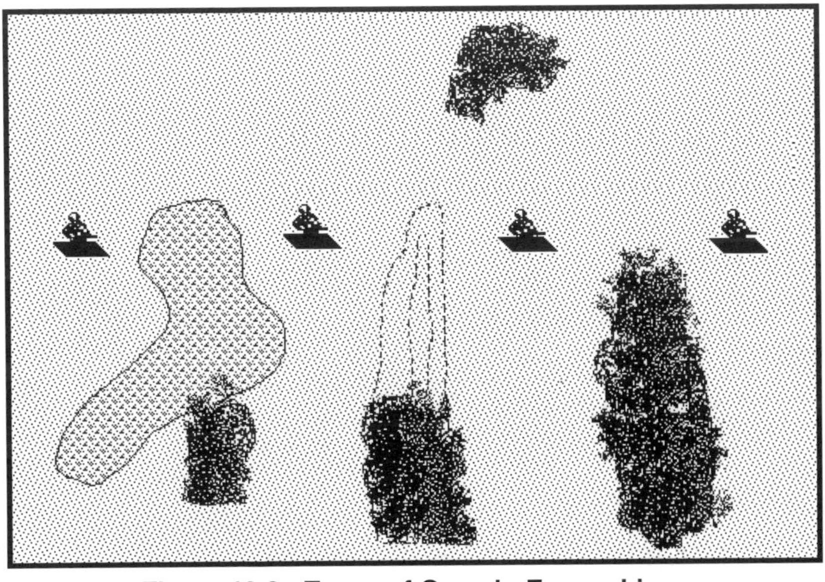

Figure 19.2: Types of Gaps in Enemy Lines
(Sources: FM 7-11B3 [1976], p. 2-VII-C-4.4; FM 7-70 [1986], p. 4-20; FM 5-103 [1985], p. 4-6.)

Figure 19.3: Each Parallel Lane Has a Pre-Confirmed Gap
(Source: FM 7-11B1/2 [1978], p. 2-II-B-3.3.)

Figure 19.4: A Single Buddy Team per Lane
(Source: FM 22-100 [1983], p. 36.)

Figure 19.5: Lead Men Move Toward Their Assigned Route
(Source: MCRP 3-02H [1999], fig. I-2.)

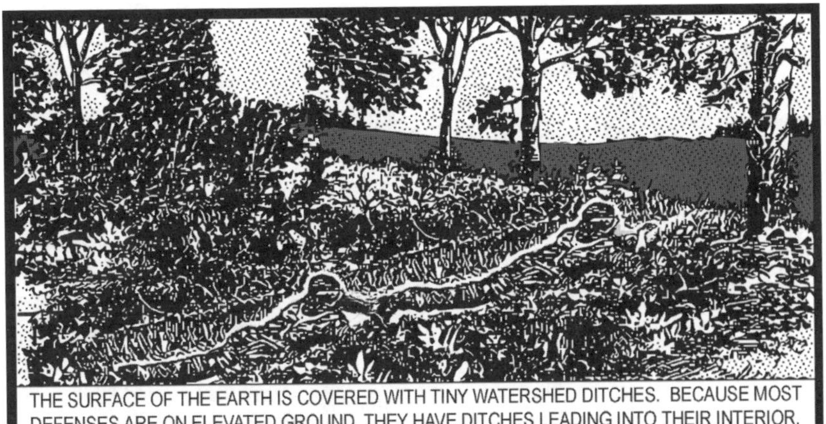

THE SURFACE OF THE EARTH IS COVERED WITH TINY WATERSHED DITCHES. BECAUSE MOST DEFENSES ARE ON ELEVATED GROUND, THEY HAVE DITCHES LEADING INTO THEIR INTERIOR.

Figure 19.6: One Team Follows a Watershed Ditch
(Source: FM 5-103 [1985], p. 4-4.)

Figure 19.7: Adjacent Team Enjoys Best Infiltration Route
(Source: FM 21-76 [1970], p. 43.)

Figure 19.8: Third Must Mask Noise of Moving Through Brush
(Source: "Podgotovka Razvegchika: Sistema Spetsnaza GRU," © 1998 by A.E.Taras and F.D. Zaruz, pp. 144, 145, 152.)

Figure 19.9: Clearings Take Camouflage and Very Slow Movement
(Sources: FMFM 0-1 [1979], p. 7-7; FM 21-75 [1967] p. 2-II-A-4.2.)

Figure 19.10: Foe Becomes Visible Against Lighter Background
(Source: FM 19-95B/CM [1978], cover; MCRP 3-02H [1999], fig. I-3.)

Figure 19.11: Stalker Keeps Object Between Himself and Sentry
(Source: "Podgotovka Razvegchika: Sistema Spetsnaza GRU," © 1998 by A.E. Taras and F.D. Zanuz, p. 373.)

Figure 19.12: Sapper Team Approaches Jump Off Point
(Source: FM 5-103 [1985], pp. 5-2, D-9.)

Figure 19.13: Lead Man Visualizes Move While Studying Gap
(Source: "Podgotovka Razvegchika: Sistema Spetsnaza GRU," © 1998 by A.E. Taras and F.D. Zanuz, p. 373.)

Figure 19.14: Forward Member Departs After Sentry Distracted
(Source: MCRP 3-02H [1999], fig. I-2.)

Figure 19.15: Backup Man Draws "Bead" on Sentry's Head
(Sources: FM 7-11B/C/CM [1979], p. 2-5; FM 5-20 [1968], p. 31.)

Figure 19.16: Penetration Occurs During a Diversion
(Sources: FMFM1-3B [1981], p. 4-9; FM 5-103 [1985], p. 4-4.)

Figure 19.17: If That Sentry Gets Nervous
(Source: FM 17-1 [1966], p. 179.)

What Makes This Type of Attack Less Risky

There are any number of naturally occurring gaps in enemy lines: (1) too large a distance between holes; (2) any swampy area; (3) a bisecting watershed ditch; (4) a fallen log; (5) a lengthy tree shadow; (6) uncleared brush; (7) an intervening mound; (8) etc. By itself, that gap doesn't permit a successful passage. But, when combined with some distraction or inhibition of a sentry, it will. Confounding his senses may be a cigarette puff, cloud over the moon, far-away flares, passing helicopter, moving armor, wind, rain, distant shelling, watch change, line check, or insect bite. Then limiting his response might be dreaming of home, lack of confidence, not wanting to give away his position through shooting, hesitancy to wake a superior, or no personal experience as a sapper.

Instead of forcing one's way into an enemy camp, the short-range infiltrator obeys an Oriental principle: "Wait for nature and other circumstances to provide an opportunity." By reaching the assigned gap around 1:00 A.M., each GI team would have a couple of hours to study it and await a break in the sentries' alertness. Then, the remaining two dark hours are plenty of time for rendezvous, explosive rigging, and exfiltration.

As the Yank buddy team members take turns passing through enemy lines, each draws a bead on the nearest sentry's head. But, that sentry need not be killed. His demise would forfeit a good exit route. After sneaking by a fully awake human being from the front, passing an unconscious one from the back should be easy. And, with no evidence of frontline entry, the foe camp commander is more likely to chalk up all damage to a mechanical breakdown or lucky mortar hit. Then, the same entry technique can be used over and over again.

Every tiny team of a short-range-infiltration has the power to complete the entire mission by itself. If a tank, command bunker, or ammunition dump is the target, each team carries enough explosive with a timer (to permit pre-detonation escape). That means fewer people ultimately at risk. In the attack just imagined, only one of the squad's six teams must like the situation well enough to attempt a penetration. The other five must only remain in the vicinity to protect that lone entrant's escape.

A dozen wartime draftees took part in this attack, and only the most confident pair had to close with the enemy. That's a lot less dangerous than hundreds of "lifers" rushing the dug-in machineguns that will invariably survive the bombardment of an enemy strongpoint.

Ambushing Can Be Equally Safe

Through claymore mines (and no small-arms fire), ambuscades can be made to look to their intended victim like a mine field or mortar attack. This mistaken impression protects detonating personnel who quickly reel in the wires from a human-wave attack.

Through pre-positioned explosives, tiny outposts can remotely engage opposition scouts and thereby preempt a full-scale attack. That may be how tiny ambushes in series became the defensive standard for UW.

Part Four

Reshuffling the Priorities a Little

"When one treats people with benevolence, justice, and righteousness, and reposes confidence in them, the army will be united in mind and all will be happy to serve their leaders."
– Sun Tzu

(Source: Attributed to Sun Tzu.)

20 Washington's Perspective

- Why are governmental perceptions so often skewed?
- Can an ongoing war be adversely affected by this?

At the highest point of a top-down society.

(Source: Copyright-free cliipart from clipartlibrary.com, #1625821, image designator "pTqKXj6jc.jpg.")

Capital City Creates America's Toughest Environment

The multi-party system of governing is the strongest in the world. But, polarity can develop between opposing coalitions, and members of Congress refuse to recognize opposing-platform strengths. When not enough intra-party negotiations take place, very little that is newsworthy escapes Washington without being somehow spun.

A capitalistic economy is also better than its socialistic counterpart. But within either, some governmental agencies may be forced to accept unwanted goods and services.

Trust in the U.S. Military

Since the early days of America, the U.S. government has closely controlled the political power of its war department. The army's commander in chief has been the President, and no service branch allocated many funds to spend as it wished. In turn, the U.S. military has maintained strict discipline over its lengthy chain of command. Subordinates at every level were expected to do exactly as told. Unfortunately, ground wars can no longer be won that way. The bottom echelons must be allowed enough leeway to deal with frontline developments.

How Things Normally Work in Washington

The administration of America has been divided between its various agencies—the most important of which may be the State Department. Whenever it fails to explain U.S. foreign policy, another war is possible.

Unfortunately, each embassy has a political appointee in charge who may make little attempt to understand local problems. Many of the most important embassies now look more like forts than *gratis* quarters in a foreign land. The public-relations parties, which one might expect, seldom occur.

Current Extent of Military Progress in America

By following societal and bureaucratic models that are both top-down, the U.S. Pentagon hasn't changed much over the last few wars. It still does "business as usual." Meanwhile, Asia's bottom-up way of solving problems has allowed more authoritarian nations to better assimilate battlefield lessons.

The Curse Is Cultural

There are two predominant lifestyles (or cultures) in this world—Eastern and Western. The first is older than the second. The bigger populations east of Turkey have learned to solve more of their problems from the bottom up through collective effort. That's how thousands of coolies with axes and shovels—instead of bulldozers—secretly built a supply road into Dien Bien Phu in early 1954.

Then, possibly due to illiteracy, the Western (or mostly European) societies allowed their leaders to make major decisions without much citizen input—i.e., from the top down. After a while, Western perspective became that of the political party in power. And within the U.S. military, the procedures for implementing strategy fell into general categories. And where overwhelming firepower was available, the tactical possibilities for each combat situation seemed no longer important.

Cultural inadequacies are hard for a proud people to admit. U.S. military commanders have, on occasion, bypassed subordinate leaders to find out what really happens where the rubber meets the road. The troops have grumbled about how dangerous "high-diddle-diddle-up-the-middle" can be. But, those troops still had to pursue most strategies through established procedure. Only on occasion, could they innovate. At Concord, for example, a chance encounter was handled through semi-autonomous elements cooperating. But ever since, the American military has allowed very little bottom-up influence on how it fights.

Most Wartime Strategies Are Established in Washington

During the Iraq War, there was a useful precedent of sorts. A fine U.S. general named Petreaus who later fell into disfavor at the CIA, did something quite unusual for a field commander. He solicited the advice of outside sources on how to conduct the war. Via e-mail to an aide, Posterity Press suggested compartmentalizing Baghdad. After others had likely said the same thing, that strategy then helped to make the Surge a success. Among that general's collocated advisors had been a well-published Australian Army officer by the name of David Kilcullen.

On 2 April 2015, Lt.Col. Kilcullen recommended a successor to the standard Western assault formation through the *Australian Army Journal*. He proposed assigning a fire support role to most of the assault force, and then allowing small elements to leapfrog forward (instead of on line). This alternate way of closing with the enemy involved seizing several pieces of strategically important ground from which opposition fire could be further suppressed. Most had probably been defensive strongpoints with their main gun recently put out of action. This can be concluded from the word "clear" in this article excerpt: "they [some infantry elements] move from point to point on a cycle of observe-suppress-move-clear-observe."

While such a progression makes more sense than the traditional on-line assault, it still involves an Attrition Warfare version of "recon pull." The German Stormtrooper squads of 1918 had been searching for weak points in the Doughboys' lines, not automatic-weapon sites to attack from the front.

Going directly at an enemy strength for any reason is tantamount to the Gen. Robert E. Lee's miscalculation on the last day at Gettysburg. As Pickett's skirmish lines were picked apart crossing the wheatfield, surviving members headed toward the copse of trees at the left end of Cemetery Ridge. Gen. Lee so admired his troops' courage that he had indirectly sent them into the belly of the beast.

Though not as dangerous, such a move was characteristically Western. In the quest for timely compliance, America's modern-day maneuvers still lack the patience to routinely succeed. At Gettysburg, the sudden reduction of Union cannon fire hadn't meant any shortage of ammunition. And the rest is history. A better choice for Gen. Lee would have been to wait for a moonless portion of the next night to negotiate that awful field.

In effect, a top-down Western army has no faith in the looser controls of a bottom-up Eastern counterpart. It wants all units doing precisely as told as soon as possible. Within its manpower pool has been instilled unquestioning loyalty to those in charge. A false sense of security ensues. The junior American infantryman concludes that his best chances of survival lie in closely following his field training, tactics manuals, and immediate superiors' directives.

But, Kilcullen's idea—though more effective than the standard U.S. assault—is still the product of 2GW logic. He is trying to bring order to something that—by its very nature—is chaotic. As at Peleliu, Iwo Jima, and Okinawa during WWII, Western forces have little chance of gaining fire superiority over a highly deceptive and fully dug-in Eastern adversary. Then, with the best of intentions, the Australian Lt.Col. says something quite controversial. He says that suppression—rather than maneuver—leads to victory in close combat.

> Maneuver is still critically important, but it happens before, after and around the flanks of the close battle; in the close battle itself, suppression is the key to success.[1]
> — *Australian Army Journal*, vol. 1, no., 1, 2015

In WWI, the German Stormtroopers had looked for gaps in Doughboy lines, because they wished to attack Allied machineguns from a flank. Once inside a U.S. defensive formation, such guns could be more safely approached. So, why subject U.S. troops to frontal machinegun fire when not absolutely necessary? During the WWII assault on Okinawa's Crescent Hill, and again at the Vietnam Battle of Dai Do, tiny U.S. Marine elements of 2/4 intentionally moved to one side of any automatic weapon to fire down its feeder trench.[2]

To openly suggest that maneuver is no longer necessary in close combat flies in the face of not only the martial arts, but also microterrain appreciation. To keep from getting machinegunned at short range, the assault member has only to flop down on the ground and crawl forward through its available folds. The best gunner in the world can miss a partially enfiladed and still moving target. Following Kilcullen's assessment, no Western squad leader would want to attempt the penetrations in depth of MW. Nor will any U.S. rifleman think it his prerogative to dodge enemy fire. Such things require a 3GW mindset.

By carefully studying the Eastern thought process, Posterity Press has learned the differences in perception. And such insight has led to few repercussions. By too well understanding the foe, that press may seem unpatriotic to some Western critics. But, that's a small price to pay for helping America to win another war. For a while in WWI, not every German Army officer believed in the Stormtrooper tactics that would prove so successful in the Spring Offensive of 1918.

Why Pentagon Can No Longer Decisively Win Overseas

It takes self-sufficient (or light) infantry squads to consolidate a huge swath of captured territory. Continually forcing the latest hardware on one's ground forces creates firepower-dependent (or heavy) infantry. That's why the Marines so seldom modernized their rifle company equipment.

The military/industrial complex of which President Eisenhower warned is still alive and well. One has trouble making money and winning wars at the same time. All of the "ism's" have shortfalls. According to Pope John Paul II, capitalism that is not properly held in check may self-destruct.[3] That's why he preferred the term "free-market society."

21 Battlefield Misconceptions

- Are all foreign conflicts easy for the Pentagon to wage?
- What makes them so hard to win?

29th Infantry Division in first wave ashore at Omaha Beach.

(Source: "Blue and Gray," by Ken Riley, U.S. Army National Guard Heritage Painting Series, from this url: http://battlefields.org/learn/articles/shades-blue-and-gray.)

Martial Warfare Should Be the Last Resort

In total war, there are no absolute winners. Everyone gets hurt to some extent—often by their own munitions. As pointed out by U.S. publication *FMFM 1,* any country wishing to follow 2GW parameters (directly attacking enemy defenders), must itself expect many casualties.[1] That means its wartime endeavors may not be as glorious; its professional warriors not as fulfilled; and any open combat not as decisive. If the style of war is too firepower oriented, the population to be liberated may suffer the same fate as its infrastructure.

The All-Important Combat Situation

To solve any complicated problem, one must pay close attention to every detail, however small. But, the U.S. Military feels no real need to do so. It has such a huge edge in technology and ordnance that it need not worry about any battlefield particular. Its invasion plans will fully address the enemy's order of battle and likely weaponry; yet the rank and file get little preview of what to expect in terms of terrain, enemy tactics, or man-made obstacles. U.S. assault troops in Vietnam were almost never warned of minefields or barbed wire around an objective. Then, lacking wire cutters or grappling hooks, they went charging up some contested hill as if fully prepared defensive positions were a thing of the past. In effect, the reconnaissance required of a deliberate assault had never occurred. Meanwhile, the NVA/VC routinely reconnoitered their assault objectives from the inside.[2]

U.S. Firepower Had Not Lived Up to Expectations

There are now official statistics to show that massive firepower made little difference in the last battle of the Pacific at Okinawa,[3] or in Vietnam.[4] Why the Pentagon has yet to acknowledge this fact is anyone's guess, but the American public needs to know. Aircraft carriers can no longer defend themselves because their accompanying ships don't carry enough anti-aircraft ammunition to stop a concerted missile attack. And the grunts must become more tactically adept to disperse under the threat of killer drones.

The World Looks Different from Ground Level

At present, commissioned members of both U.S. infantry branches partake in endless mental gymnastics on how finally to practice their new 3GW doctrines. In a highly stratified bureaucracy running mostly on well-established procedure, such discussions may create the impression of necessary reform. But, within in the average "professional journal" article, lies the truth. It may make sense to a high-tech staff officer, but holds little of value for the fledgling brown bar about to lead 40 young grunts into mortal combat.

Top-down thinkers will thoroughly explore every level of information on their way down to the targeted issue. So doing becomes such a monumental task they get mired in an inane discussion that is more about terminologies than solutions. In essence, *Occam's Razor* has been badly violated, or—in enlisted lingo—the KISS (keep it simple stupid) principle.

When one looks at a wartime dilemma from the bottom-up, he first considers the opposing capabilities—the micro-basics with which America's under-armed foes have generated so much surprise. With regard to the U.S. basic of "move," a good example would be how to cross swampy ground without a sound. Then, any unique aspects of the microterrain are looked at (before the battlefield as a whole). What results is a better way to escape the effects of enemy weaponry. Not only has the investigation been less broad, but better focused on situational detail. If recent history is any indicator, the prize for shifting to the Eastern thought process would be fewer friendly casualties and more conflicts decisively won. The latter takes consolidating an immense area—something only possible through tiny, self-sufficient outposts.

At some point, the core issue must be faced. It's not particularly complicated, nor will it please the American public. The U.S. infantry branches can't fully practice their new 3GW doctrine, because they're not willing to decentralize control over training and operations. MW requires tiny elements so tactically adept as to function as the point of an offensive spear or building block of a defensive matrix. Such capabilities are most easily achieved from the bottom up through field testing collective NCO recommendations. Then, as in football, all resulting techniques must be frequently practiced.

Nor would a "military/industrial complex" see any need for its predominantly mechanized ground forces to embrace the micro-basics of dismounted warfare. Expert light infantrymen need no fancy devices to hold their own in modern combat. That's something that profit-oriented arms manufacturers recommend.

So, Which Side Has Better Tactics?

Good small-unit tactics are those that produce the fewest friendly casualties. The most logical way to achieve them is to let the people at most risk provide some input. That's the riflemen themselves. But, a rifleman can't do whatever he wants while part of a parent-unit maneuver. So, such influence must be levied during the training cycle. A few individualized or more surprise-oriented techniques must be added to the rifleman's current portfolio of standardized (and thus fully predictable) procedures. Even then, he will likely be told when to use them. Only when facing the foe alone will he have more leeway. Otherwise, a well-meaning headquarters may find its small unit advice still following pre-machinegun parameters in 2024.

If the Role of the Officer Is Still to Lead

To successfully communicate with enlisted personnel, a U.S. Marine officer must talk much differently than he has been lately writing in the *Gazette*. His instructions must be short, to the point, and clear on how much leeway exists in their execution. Often, waiting for dark would make them easier to perform. Initially, most orders lack enough tactical detail. Unless some subordinate adds it, too many casualties will result. For example, a sapper attack more safely handles a foe tank than full-blown assault.

The Enlisted Path to Glory

Cpl. Herschel Woody Williams was a 21st Marines demolition sergeant on Iwo who repeatedly went forward alone to destroy foe pillboxes. This effort had been complicated by first borrowing a flamethrower. That's like bathing in gasoline before trying to light a campfire. When asked at the age of 98 what had motivated his MoH actions, Williams replied that "the cause had been greater than he was."[5] The average enlisted grunt on Iwo may have shared this opinion. Or, he had learned from some Western movie "that a real man made sure his death amounted to something."[6] Either way, "uncommon valor was soon a common virtue." How Iwo's three strongpoint belts were finally penetrated was hard for U.S. COs to watch. They would have tried something less casualty producing, but were hesitant to alter what had just begun to work. Near the end of the battle, the 3rd Marine Division commander remarked that the battle's outcome was never in doubt, only whether any troops would be left to see it.[7]

22 The American Rifleman's Experience

- How trusting of their service branch should U.S. recruits be?
- Will that help them to survive in battle?

Young American grunt not allowed the "counsel of his fears."

(Source: "Under Fire," by SFC Darrold Peters, from U.S. Army "War on Terror Images," at this url: http://www.history.army.mil/books/wot_artwork/images/52a.jpg.)

Life at the Bottom of the Heap

Of late, the U.S. military has become far too officer-centric. In the Basic School instructor hall at Quantico hangs a quote from a 19th Century admiral—"officers win wars." The job of a Marine officer isn't easy; many officers have been killed leading from the front; and every officer fills a key role in mission accomplishment. But, few of the enlisted veterans of America's many wars would give credit for the bunker they just took to a commanding officer. And, as this book has again shown, most of those conflicts were won one bunker at a time. Large U.S. units have conducted feints during opposed landings, but lower echelons couldn't generate much surprise ashore. In fact, the early WWII version of "blind 'em, burn 'em, and blast 'em" entailed no deception whatsoever.

All Marine support personnel largely exist to help the infantry. Rear-area Army units are told this as well. But, a few non-infantry officers still have difficulty relating to junior riflemen. For a while, the Chinese Communists solved this problem by turning PRC staff officers into infantry Privates six months out of every year. But, the USMC has only occasionally idolized its junior infantrymen: (1) with the "Rifleman's Creed" from Gen. Rupertus; and (2) *A Message to Garcia* on required reading lists. For a while, the U.S. Army was "an Army of One." Famous generals had praised entry level fighters. Patton talked about how the troops had a nose for what was true. And a former head of 4th Marine Division referred to himself as Pvt. Barker (after only being an officer candidate).[1] In truth, the junior members of a military bureaucracy may be more important to final victory than those at the top.

It Takes Confident Soldiers to Win Wars

Are America's youth given enough leeway to deal with all the circumstances of combat, or must they follow instructions ignoring a few. In any war not decisively won, every death will be in vain. Thus at issue is whether each U.S. soldier's life is being properly valued.

Almost every improvement to small-unit tactics has been attempted by Western units at some point in their history—many by native Americans. So, GIs need not copy what Eastern counterparts develop. They must only dig far enough into their backgrounds for an appropriate procedure. The continual flow of new equipment makes such searches difficult. Stand-off ordnance, for example, seems to lessen the need for close-range skills.

What Constitutes a Sophisticated Maneuver

The last evolutionary step in squad tactics occurred in 1917 with German soldiers attacking a fully prepared defense line by noisy surprise. They did so through a fake artillery/mortar attack involving concussion grenades and shrapnel-free satchel charges.

But, East Asians have offensively applied all sorts of ruses and diversions since Sun Tzu (544 to 496 B.C.) first announced that "all warfare is based on deception." Only within England, France, and America have overwhelming firepower seemingly eliminated the need to fool one's foe during a ground assault. Charging in from a direction other than one's base of fire won't be enough of a diversion.

Machineguns had led to the change in squad tactics of 1917, but played no role in the Stormtroopers' illusion. Recent inventions may have a lesser effect on modern-day maneuver, yet take no part in the countermeasure. The saving procedure will be of a light-infantry nature, involving micro-basics. For example, one most easily defeats a heat-seeking device by remaining in defilade.

The evolution of squad tactics has been based more on frontline sense than technological progress. So, how to supersede any new gadget will normally exist in the historical record.

Private First Class (PFC) Sledge's Descriptions

The average U.S. infantryman is like a big-game hunter. He lives outdoors, seeks dangerous prey, and constantly risks being killed by friendly fire. That's why he "stays off the skyline," as was the 6th Marine Division's motto on Okinawa.[2] Though often scared, that young Yank doesn't consider his job heroic. He wants no formal recognition, only to return home alive. So doing requires the proper combination of personal equipment, individual movement, and headquarters leeway. Yet, he has little control over any of the three.

While the number of U.S. riflemen who have recorded their wartime experiences is limited, there does exist a 60mm mortarman who survived many of the same circumstances during two of the fiercest battles of WWII. His name was Eugene Sledge.

On Peleliu, 1st Marine Division were forced to traverse a huge airfield clearing on line in September of 1944. Midway across, it ran into a long-range firestorm from automatic weapons on both flanks and artillery atop the Umurbrogol. Those hundreds of Marines were on a flat coral surface, with only a few shallow shell holes. When they realized a huge ambush was underway, they had no choice but to move a little faster.

> We moved at a walk and then a trot, in widely dispersed waves. Four infantry battalions—from left to right 2/1, 1/5/2/5, and 3/5 . . . moved across the open, fire-swept airfield.[3]
>
> — *With the Old Breed,* by Sledge (of 3/5)

Even with this increased tempo of transit, the heavily equipped young Americans remained sadly vulnerable to the spray of killer shards.

> We moved rapidly in the open . . . . through ever increasing enemy fire. I saw men to my right and left running bent as low as possible. . . . In many respects it was more terrifying than the landing. . . . We were exposed, running on our own power through a veritable shower of deadly metal. . . .
>
> . . . Japanese bullets snapped and cracked, and tracers went by me on both sides at waist height. This deadly small-arms fire seemed almost insignificant among the erupting shells. . . .
>
> . . . How far we had come in the open I never knew, but it must have been several hundred yards. . . .
>
> . . . The attack across Peleliu's airfield was the worst combat experience I had during the entire war. It surpassed . . . all the subsequent horrifying ordeals on Peleliu and Okinawa.[4]
>
> — *With the Old Breed,* by Sledge (of 3/5)

Then, as luck would have it, PFC Sledge would get to endure the 82 day nightmare of Okinawa. During one episode, he and his buddies were forced to shuttle their heavy boxes of ammunition across a fire-swept draw screened by smoke. Except for the smoke, Eugene's experience then mirrored what a rifleman routinely endures on an assault.

> The Nambu kept firing down the smoke-covered draw. I was reluctant to go, as were the others, but we could see Redifer standing out in the draw, throwing more phosphorous grenades to hide us. I felt like a coward. . . . Someone said resignedly, "Let's go, on the double and keep your five-pace interval."
>
> We dashed into the smoky, murky air. I lowered my head and gritted my teeth as the machinegun slugs

> snapped and zipped around us. I expected to get hit. . . . I wasn't being brave, but Redifer was, and I would rather take my chances than be yellow in the face of his risks to screen us.[5]
>
> — *With the Old Breed,* by Sledge (of 3/5)

Until the Pentagon embraces the long-standing state of the art for small-unit infantry tactics, young U.S. grunts will face the same peril. So, instead of the modern-day "thanks for your service," "lest we forget" might convey more personal commitment to national security.

Fear is a God-given instinct that helps people to stay healthy. Ignoring what makes one afraid is like disregarding common sense. A wider assortment of tactical techniques and the leeway to use them might not only increase the U.S. rifleman's degree of confidence, but also his chances of survival.

> There is a time to take counsel of your fears.[6]
>
> — George S. Patton

The Modern-Day U.S. Rifleman

The USMC's magnificent heritage is largely based on how its frontline fighters had been prepared for combat. After receiving discipline, infantry basics, and team spirit at boot camp, they were sent to their field unit to learn the finer points of mission accomplishment. In between had been another school at which introductory 2GW procedures were covered, but not when more personal initiative or micro-basics might be authorized. The U.S. Army regimen was similar.

Many GIs then entered the melee highly exhilarated by the prospect of demonstrating more courage than previously available. But, a closer look at the battles of Iwo Jima, Peleliu, and Okinawa show a different frontline attitude would have better facilitated the final victory. The still standing U.S. Private was a student of the situation, of all those little things headquarters had forgotten to mention. He had slightly modified all battlefield behavior to account for those hidden dangers. While making no attempt to be brave, he focused on safely removing the next impediment. He knew that backing up is sometimes necessary to move forward somewhere else. And, he worried more about fellow team members than himself. In other words, he was no longer an impending casualty, but rather a Tokyo flag raiser. Largely through his own efforts, he had become a long-term player—that which the average commander would have much preferred to begin with.

Today's U.S. rifleman continues to be trained in 2GW, though most enemy counterparts are now 3GW qualified. So, that young GI will likely be a "flash in the pan" during any direct enemy encounter. Yet, someone with more combat agility might last long enough to help with the final battle. It is the American standard—high-tech 2GW—that demands strict obedience to all headquarter directives (however devoid of detail). Moving into the more productive 3GW requires obeying only the spirit of one's orders. And, absolute loyalty to one's command and Corps gives way to more support for the bottom-echelon combatant.

Some say the U.S. military (as currently structured and utilized) is a fickle mistress. It's beautiful to look at, but very dangerous for a young rifleman to be around.

Handwriting on the Wall

Since WWI, America's carefully choreographed large-unit formations have been severely attrited through long-range machinegun fire. In effect, the all-powerful phalanx made too lucrative a human target. Over broken ground, sending all infantrymen forward in armored personnel carriers does no good. They lose their mastery of the microterrain. And, as recently demonstrated in Ukraine, all armor is quite vulnerable to the latest hand-held antitank weapons. Like the foe in Vietnam, dismounted U.S. "ground pounders" must widely disperse and then secretly come together for any large-unit attack. That takes decentralized control and more light-infantry expertise. Every U.S. squad must be tactically adept enough to cross enemy territory alone without any supporting-arms umbrella. Otherwise, no degree of GI sacrifice will win the next big war.

Notes

SOURCE NOTES

Illustrations:

Cover art is a WWII recruiting poster entitled "Americans will always fight for liberty," by Bernard Perlin. It was painted for the OFFICE OF WAR INFORMATION in Washington, D.C., around 1943-1945. As such, it is now in the public domain and currently part of the National Archives in College Park, MD, with identifier 513806. Made available for reprinting through Wikimedia Commons, the file has been lightened, reversed, and cropped but left otherwise unchanged from the original. It was retrieved in August 2022 from the following url: https://commons.wikimedia.org/wiki/File: Americans_will_always_fight_for_liberty -_NARA_-_ 513806.jpg.

Picture on page 3 is from U.S. AIR FORCE (www.af.mil). This graphic was likely drawn by a paid member of the U.S. Armed Forces and is thus in the public domain. With image designator, "AFG-060725-004 .jpg," it was retrieved in early 2006 from the following url: http://www.af.mil/ shared/media/ggallery/hires/afg_030409_002.jpg.

Images on pages 25, 71, 155, 161, and 265 are reprinted from NATIONAL GUARD IMAGE GALLERY (http://www.nationalguard. mil/resources/photo_gallery/heritage/index.htm) with designators: "arramadi.jpg," "Plenty of Fighting Today," "cornrowtohedgerow.jpg," "sunshinedivision.jpg," and "Blue and Grey," respectively. Only for an educational nonprofit is that fair usage. Others must be more mindful of possible copyrights.

Images on pages 26, 33, 35, 36, 55, 57, 62, 80, 99, 101, 109, 113, 118, 120, 128, 130, 135, and 138 are from WIKIMEDIA COMMONS and all in the public domain. See each picture's separate caption for the appropriate image designator.

Picture on page 38 comes from a painting of Rogers Rangers, "To Range the Woods, New York, 1760," by Army Artist Team XXII, is in the public domain from U.S. ARMY CENTER OF MILITARY HISTORY at the following url: http://www.history.army.mil/html/artphoto/pripos/prponco.html.

Pictures on pages 115, 143, and 169 are reprinted from "Artphoto Archives" of the U.S. ARMY'S CENTER OF MILITARY HISTORY. All are considered to be in the public domain. They have these illustration designators: "avop05-98_1.jpg" "avop05-99_1.jpg," and "0507-3.jpg," respectively.

Pictures on pages 189 and 269 are public domain material from "War on Terror Images" of the U.S. ARMY at this url: http://history.army.mil/html/bookshelves/resmat/GWOT/art_and_photos.html. They have the following image designators: "19b.jpg" and "52a.jpg."

Picture on page 193 has been reproduced after written assurance from Orion Books, UK, that the copyright holders for *UNIFORMS OF THE ELITE FORCES,* with text by Leroy Thompson and color plates by Michael Chappell, could no longer be reached. It constitutes Plate 26 (No. 77) of the Orion publication. Copyright © 1982 by Blandford Books Ltd. All rights reserved.

Image on page 201 is thought to be in the public domain from the archives of U.S. ARMY CENTER OF MILITARY HISTORY ARTWORK (http://www.history.army.mil/html/artphoto/artwork.html). Its designator is: "Image-12.jpg." See picture caption for title, artist, and exact url.

Image on page 217 is from PIXY.ORG #1625821. For use by a educational nonprofit, it is deemed to be the public domain.

Picture on page 223 is reproduced with permission of Dr. Anatol Taras, Minsk, Belarus, from *PODGOTOVKA RAZVEGCHIKA: SISTEMA SPETSNAZA GRU,* by A.E. Taras and F.D. Zaruz. This illustration is from pp. 374/375 of that publication. Copyright © 1998 by A.E. Taras and F.D. Zaruz. All rights reserved.

Picture on page 238 is reproduced with permission of Raytheon and the publishers of *ARMED FORCES JOURNAL INTERNATIONAL,* from "Revolutionary Soldier," by Glen W. Goodman, Jr., October 1999, page 64. Copyright © 1999 Raytheon. All rights reserved.

Picture on page 241 has been reprinted with permission of SORMAN INFORMATION AND MEDIA and SWEDISH ARMED FORCES, from *Soldf: Soldaten i Falt,* by Forsvarsmakten, with illustrations by Wolfgang Bartsch. This illustration is from page 23, and 225, or 370 of the Swedish publication.

Image on page 259 is from CLIPART LIBRARY #963190. For use by an educational nonprofit, it is deemed to be in the public domain

Text:

Excerpt on pages 60-63 comes from "Overview of the American Revolutionary War," by the AMERICAN BATTLEFIELD TRUST, at the following url: https://www.battlefields.org/learn/articles/overview-american-revolutionary-war?ms=googlegrant&ms=googlegrant. As the associated webpage contains no copyright notice, it's assumed that this summary is in the public domain. If not, its reprinting is fair usage by an educational nonprofit.

Excerpt on pages 127 and 128 reproduced after being unable to contact the copyright holder of *THEIR TIME IN HELL: THE 4TH MARINE BRIGADE AT BELLEAU WOOD,* by George B. Clark (Pike, NH: The Brass Hat, 1996). It's from pages 40-66 of that publication. For an educational nonprofit, its reproduction is thought to be fair usage. All rights reserved.

ENDNOTES

Introduction

1. "Warfighting," *MCDP 1* (Washington, D.C.: HQMC, 1997), pp. 42, 43.

Acknowledgments

1. Pope John Paul II, *Crossing the Threshold of Hope* (New York: Alfred A. Knopf, 1995), pp. 205, 206.

Chapter 1: *The Dangerous Illusion*

1. Mao, as quoted in *China's Battle for Korea: The 1951 Spring Offensive, by* Xiaobing Li (Bloomington, IN: Indiana Univ. Press, 2014), pp. xxv-xxvi.
2. John Wukovits, *American Commando* (New York: New American Library, 2009), p. 12; "Secret Manual Gives Glimpse of North Korean Military Tactics," by Steve Herman, *Voice of America* on line, 18 September 2010.
3. Gen. Vo Nguyen Giap, "Once Again We Will Win," in *The Military Art of People's War,* ed. Russel Stetler (New York: "Monthly Review Press," 1970), pp. 264, 265.
4. *The Iwo Alamo,* by H. John Poole (Emerald Isle, NC: Posterity Press, 2020), pp. 223, 224; *Peleliu Progress*, by Poole (Emerald Isle, NC: Posterity Press, 2021), p. 216; *Okinawan Reckoning,* by H. John Poole (Emerald Isle, NC: Posterity Press, 2022), p. 28.
5. *North Korea's Hidden Assets,* by H. John Poole (Emerald Isle, NC: Posterity Press, 2018), chapt. 2.
6. "Across the Reef: The Marine Assault of Tarawa," by Col. Joseph H. Alexander, USMC (Ret.), *Marines in WWII Commemorative Series* (Washington, D.C.: Marine Corps Historical Ctr., 1993).
7. *Okinawa: The Last Battle,* by Roy E. Appleman, James M. Burns, Russell A. Gugeler, and John Stevens, *United States Army in World War II Series* (Washington, D.C.: U.S. Army's Ctr. of Mil. Hist., 2000), p. 411.
8. Memorandum for the record by H.J. Poole, after having his own M-16 jam during a firefight in Vietnam on 18 August 1968.
9. "Overstressing Safety Is Leaving Momentum Behind," by Lt.Col. H.J. Poole USMC (Ret.), *Army Magazine,* August 2020, p. 18.
10. *Peleliu Progress,* by H. John Poole (Emerald Isle, NC: Posterity Press, 2021), pp. 81-84.

Chapter 2: *It's Harder to Fight a Thinking Opponent*

1. Dictionary.com, s.v. "warfare."
2. *Wikipedia Encyclopedia,* s.v. "U.S. Enlistment Contract."
3. *Mao Tse-tung: An Anthology of His Writings,* ed. Anne Fremantle (New York: Mentor, 1962), p. 69.
4. Terrence Maitland and Peter McInerney, *Vietnam Experience: A Contagion of War* (Boston, MA: Boston Publishing, 1968), p. 97.
5. *Wikipedia Encyclopedia,* s.v. "Battles of Lexington and Concord."

Chapter 3: *What's a Tactical Technique?*

1. *Mao Tse-tung: An Anthology of His Writings,* ed. Fremantle, p. 82.
2. Truong Chinh, *Primer for Revolt,* intro. Bernard B. Fall (New York: Praeger, 1963), pp. 114-117.
3. Maitland and McInerney, *Vietnam Experience*, pp. 100, 101.
4. Col. Wesley L. Fox, *Marine Rifleman: Forty-Three Years in the Corps* (Dulles, VA: Brassey's, 2002), pp. 237, 238.
5. "Jack Lucas and the Battle of Iwo Jima," Youtube video presentation
6. Bill Rose, as quoted in "A Season In Hell: Iwo Jima 50th Anniversary," by Nathan Miller (N.p: n.d.).
7. *Super-Squad: The Now Missing Component,* by H. John Poole (Emerald Isle, NC: Posterity Press, 2019), pp. 68-70, chapt. 17.

Chapter 4: *The Need for More Small-Unit Maneuver*

1. Lt.Col. Tim Stoy U.S.Army (Ret.), review of *Okinawan Reckoning,* by H, John Poole (Emerald Isle, NC: Posterity Press, 2022), "Army Magazine," January 2023, p. 70.
2. "Crumbling ruins of the Mound of the Dead: World's only Bronze Age city could vanish in just 20 years," by Tara Brady, *The Daily Mail* (UK) on line, 14 October 2013.
3. Sun Tzu, *The Art of War.*
4. *Okinawan Reckoning,* by H, John Poole (Emerald Isle, NC: Posterity Press, 2022), chapt. 7; *The Iwo Alamo,* by Poole, p. 212, chapt. 10; *Peleliu Progress*, by Poole, chapts 8 and 15.
5. *The Iwo Alamo,* by Poole, p. 212.
6. *Peleliu Progress*, by Poole, chapt. 5.
7. Thomas M. Huber, "Japan's Battle for Okinawa, April - June 1945," *Leavenworth Papers No. 18* (Ft. Leavenworth, KS: Combat Studies Inst., U.S. Army's Cmd. and Gen. Staff College, 1990), chapt. 3.

8. Hiromichi Yahara, *The Battle for Okinawa,* trans. Roger Pineau and Masatoshi Uehara (New York: John Wiley & Sons, 1995), pp. 63, 64; "Remembering Okinawa," by Robert Akers Martin (N.p.: *Lulu.com,* 2014); *Okinawa: The Last Battle,* by Appleman et al, p. 256.

9. Maj.Gen. Vandegrift, as quoted in *Fighting on Guadalcanal, FMFRP 12-110,* from U.S.A. War Office (Washington, D.C.: U.S. Govt. Printing Office, 1942), p. v.

Chapter 5: *The French and Indian Wars*

1. *Encyclopedia Britannica,* s.v. "The Iroquois League."

2. *Frontier Skills: The Tactics and Weapons That Won the American West,* by William C. Davis (Guilford, CT: Lyons Press, 2003), p. 15.

3. Antoine de la Mothe Cadillac, as quoted in *Frontier Skills,* by Davis, p. 14.

4. *The Way of the Warrior,* editors of Time-Life Books (Alexandria, VA: Time-Life Books, 1993), pp. 129-131, in *Frontier Skills: The Tactics and Weapons That Won the American West,* by Davis, p. 15.

5. "Ottowa Indians," Ohio History Central, n.d.

6. *Encyclopedia Britannica,* s.v. "Edward Braddock."

7. *Robert Rogers' Rules for the Ranging Service: An Analysis,* by Matt Wulff (Westminster, MD: Heritage Books, 2007), p. 2.

8. *Terrorist Trail: Backtracking the Foreign Fighter,* by H. John Poole (Emerald Isle, NC: Posterity Press, 2006), chapt. 10.

9. *Robert Rogers' Rules for the Ranging Service,* by Wulff, pp. 3-5, 32, 33.

10. *Encyclopedia Britannica,* s.v. "Rogers' Rangers."

11. *Wikipedia Encyclopedia,* s.v. "Rogers' Rangers."

12. *Robert Rogers' Rules for the Ranging Service,* by Wulff, p. 93.

13. *Wikipedia Encyclopedia,* s.v. "Robert Rogers' 28 'Rules of Ranging'."

14. David Scott-Donelan, *Tactical Tracking Operations: The Essential Guide for Military and Police Trackers* (Boulder, CO: Paladin Press, 1998), p. 3; *Robert Rogers' Rules for the Ranging Service,* by Wulff, pp. 10-35.

15. *Robert Rogers' Rules for the Ranging Service,* by Wulff, pp. 37-43.

16. David Scott-Donelan, *Tactical Tracking Operations,* p. 3.

17. Ibid.

18. *Robert Rogers' Rules for the Ranging Service,* by Wulff, pp. 53-58.

19. Ibid., pp. 59-62.

20. Michael Lee Lanning and Dan Cragg, *Inside the VC and the NVA: The Real Story of North Vietnam's Armed Forces* (New York: Ivy Books, 1992), pp. 206-208.
21. *Robert Rogers' Rules for the Ranging Service,* by Wulff, pp. 65-68.
22. *Peleliu Progress,* by Poole, p. 135.
23. *Robert Rogers' Rules for the Ranging Service,* by Wulff, pp. 71-76.
24. Ibid., p. 83.
25. Ibid., p. 84.
26. Ibid., p. 33.
27. Ibid., pp. 19, 95-103.
28. Ibid., pp. 105-109.
29. *Phantom Soldier: The Enemy's Answer to U.S. Firepower,* by H. John Poole (Emerald Isle, NC: Posterity Press, 2001), chapt. 7.
30. *Robert Rogers' Rules for the Ranging Service,* by Wulff, p. 115.
31. Ibid., p. 116.
32. *Okinawan Reckoning,* by Poole, chapt. 9.
33. *Robert Rogers' Rules for the Ranging Service,* by Wulff, pp. 117-122.
34. Ibid., p. 128.
35. *The Tiger's Way: A U.S. Private's Best Chance for Survival,* by H. John Poole (Emerald Isle, NC: Posterity Press, 2003), chapt. 13.
36. *Robert Rogers' Rules for the Ranging Service,* by Wulff, pp. 131-136.
37. Ibid., p. 139.
38. Ibid., pp. 140-142.
39. Ibid., pp. 143-49.
40. Ibid., pp. 151-161.
41. Ibid., pp. 163-166.
42. Ibid., pp. 169-172.
43. David Scott-Donelan, *Tactical Tracking Operations,* p. 3.
44. *Robert Rogers' Rules for the Ranging Service,* by Wulff, pp. 181-184.
45. Ibid., pp. 187-190.
46. Ibid., pp. 191-197.
47. Ibid., pp. 199-207.
48. Ibid., pp. 209-212.
49. Ibid., pp. 213-216.
50. Ibid., pp. 219-223.
51. Hoang Khoi, *The Ho Chi Minh Trail* (Hanoi: The Gioi Publishers, 2001), p. 46.
52. Scott-Donelan, *Tactical Tracking Operations,* pp. 85-94.
53. *Combat Tracking* (Paluda, Malaysia: British Army Combat Trng, Ctr., n.d.), p. 23.

54. Allan A. Macfarlan, *Exploring the Outdoors with Indian Secrets* (Harrisburg, PA: Stackpole Books, 1971), p. 90.
55. *Robert Rogers' Rules for the Ranging Service,* by Wulff, p. 99.
56. Sun Tzu, as quoted in "Warfighting," *FMFM 1* (Washington, D.C.: HQMC, 1989), p. 55.
57. Ibid., pp. 28, 29.
58. *Robert Rogers' Rules for the Ranging Service,* by Wulff, p. 2.
59. "Warfighting," *FMFM 1* (Washington, D.C.: HQMC, 1989).
60. *Sun Bin's Art of War: World's Greatest Military Treatise,* trans. Sui Yun (Singapore: Chung Printing, 1999), p. 190.
61. Keith William Nolan, *Operation Buffalo: USMC Fight for the DMZ* (New York: Dell Publishing, 1991), p. 79.
62. Ibid., p. 77.
63. "Tactics," MCDP 1-3 (Washington, D.C.: HQMC, 1997), p. 49.
64. Stevenson summary of "War on the Run: The Epic Story of Robert Rogers and the Conquest of America's First Frontier," by John Ross, *Tactics and Preparedness*, November 2021.
65. "Warfighting," *FMFM 1,* pp. 60, 61.
66. *Okinawan Reckoning,* by Poole, chapts. 2, 3.
67. *Guardian Joe: How Less Force Helps the Warrior,* by H. John Poole (Emerald Isle, NC: Posterity Press, 2018), chapt. 9.

Chapter 6: *U.S. War of Independence*

1. Pres. John F. Kennedy, in quote from his 1962 West Point graduation speech, in "Unconventional Warfare Fundamentals," at *The Irregular Warrior* website.
2. "Native Americans in the Revolutionary War," *History of Massachusetts Blog.*
3. "Battle of Bunker Hill," *History Channel,* n.d.; *New World Encyclopedia,* s.v. "Battle of Bunker Hill."
4. *Encyclopedia Britannica,* s.v. "Battle of Kings Mountain."
5. Attributed to Sun Tzu.
6. *Gale Encyclopedia of Native American Tribes,* ed. Sharon Malinowski and Anna Sheets (New York: Gale, 1998), p. 308.
7. "Native Americans in the Revolutionary War," *History of Massachusetts Blog.*
8. Ibid.
9. "Ottawa Indians," *Ohio History Central,* n.d.; "Native Americans in the Revolutionary War."
10. "Native Americans in the Revolutionary War," *History of Massachusetts Blog.*

11. *Gale Encyclopedia of Native American Tribes,* eds. Malinowski and Sheets, p. 279.

12. *The Way of the Warrior,* eds. of Time-Life Books, pp. 129-131, in *Frontier Skills,* by Davis, p. 15.

13. "The War of the American Revolution," by Robert W. Coakley and Stetson Conn, *CMH_Pub_70-6-1* (Washington, D.C.: U.S. Army's Ctr. of Mil. Hist., 2010), p. 10.

14. Ibid., p. 23.

15. "Overview of the American Revolutionary War," *American Battlefield Trust.*

16. *New World Encyclopedia,* s.v. "Guerrilla Warfare," as retrieved in April 2022 from the following url: https://www.newworldencyclopedia.org/entry/Guerrilla_warfare.

17. *Robert Rogers' Rules for the Ranging Service,* by Wulff, p. 2.

18. "The War of the American Revolution," by Coakley and Conn, *CMH_Pub_70-6-1,* p. 22.

19. Ibid.

20. Ibid.

21. Ibid.

22. "The American Revolution," by Wilson, doctoral research (Quantico, VA: U.S. Marine Corps Research Ctr., n.d.), p. 7.

23. Ibid., p. 11.

24. B.Gen. Samuel B. Griffith, as quoted in "The American Revolution," by Wilson, p. 11.

25. "The American Revolution," by Wilson, p. 43.

26. Ibid., p. 13.

27. "The War of the American Revolution," by Coakley and Conn, *CMH_Pub_70-6-1,* pp. 68-70.

28. Ibid., pp. 119-124.

29. Ibid., pp. 125-133.

30. Attributed to Sun Tzu.

31. "The American Revolution," by Wilson, p. 22.

32. "How Alexander Hamilton's Men Surprised the Enemy at the Battle of Yorktown," by Leslie Kennedy, *History Channel,* 19 October 2021.

Chapter 7: *U.S. Civil War*

1. "American Indian Wars." *History Channel,* 17 November 2019.
2. Ibid.
3. Ibid.
4. Ibid.
5. Ibid.

6. Ibid.
7. Ibid.
8. John MacDonald, *Great Battles of the Civil War* (New York: Macmillan U.S.A. in assoc. with Marshall Editions, Ltd., 1992), p. 78.
9. "Significant Civil War Battles," *American Experience, PBS TV.*
10. Ibid.
11. "Guerrilla War in the North," *Encyclopedia.com.*
12. "Did Lt. Gen. Richard Ewell Lose the Battle of Gettysburg," *HistoryNet,* n.d.
13. Gregory Jaynes and eds. of Time-Life Books, *The Civil War: The Killing Ground—Wilderness to Cold Harbor* (Alexandria, VA: Time-Life Books, 1986), p. 33.
14. Ibid.
15. James M. McPherson, ed., and Richard Gottlief, managing ed., *Battle Chronicles of the Civil War 1864* (New York: Grey Castle Press, 1989), p. 68.
16. Noah Andre Trudeau, *The Last Citadel* (Boston, MA: Little, Brown & Co., 1991), pp. 332-341.
17. "Battle of Jackson," Civil War: Secret Missions, *History Channel,* 15 January 2022; "The Civil War Siege of Jackson, Mississippi," by Jim Woodrick, *Civil War Series* (N.p.: History Press, 18 April 2016).
18. "Discipline and Training of Soldiers," Civil War Leadership blog, 15 April 2015.
19. "The War of the American Revolution," by Coakley and Conn, *CMH_Pub_70-6-1,* p. 23.
20. *Wikipedia Encyclopedia*, s.v. "U.S. Department of War."
21. "The War Effort: The War Department," Mrlincolnswhitehouse.org.
22. "Discipline and Training of Soldiers," Civil War Leadership blog.
23. *Encyclopedia.com,* s.v. "Drill Training."
24. "Military Drills of the Civil War," *civilwar.com,* n.d.
25. "How wide and long would a column be," *Civilwartalk.com.*
26. Ibid.; "From Smooth Bore to Rifled Musket," *HistoryMaking.org;* "Small Arms of the Civil War," *American Battlefield Trust;* "The Rifle-Musket vs. The Smoothbore Musket," by Justin Stanage, Dept. of Hist. Indiana Univ., n.d.
27. "The Marine Rifle Squad," *MCRP 3-10A.4,* HQMC, August 2020.
28. "Battle Hymn of the Republic: The Story Behind the Song," the Kennedy Center.
29. Guthrie, Capt. Charles "Tag" USMC (Ret.) (enlisted participant in the battle of Cam Lo and part-time historian), in e-mail message to author of 2021.
30. *Super-Squad,* by Poole, chapt. 15.

Chapter 8: *Western Indian Wars*

1. "Indians Don't Attack at Night," *Jeff Arnold's West,* 26 June 2020.

2. *Encyclopedia.com,* s.v. "Geronimo," as retrieved in March 2022 from the following url: https://www.encyclopedia.com/people/history/north-american-indigenous-peoples-biographies/geronimo.

3. "Counting Coup," by Dennis Gaffney, *PBS's Antiques Roadshow,* 24 April 2006.

4. "Everything You Know About the Indian Wars Is Wrong," *HistoryNet,* n.d.

5. *Encyclopedia Britannica,* s.v. "Shoshone People," as retrieved in March 2022 from the following url: https://www.britannica.com/topic/Shoshone-people.

6. "These four final battles spelled the end of Comanche hegemony over the Southern Plains," by Richard A. Marini, *San Antonio Express News* on line, 12 April 2017.

7. *Wikipedia Encyclopedia,* s.v. "Bear River Massacre," as retrieved in March 2022 from the following url: https://en.wikipedia.org/wiki/Bear_River_Massacre.

8. "Indians Don't Attack at Night," *Jeff Arnold's West,* 26 June 2020.

9. "The Comanche," *Visit Big Spring Texas*, n.d.

10. *Encyclopedia Britannica,* s.v. "Comanche People," as retrieved in March 2022 from the following url: https://www.britannica.com/topic/Comanche-people.

11. "These four final battles spelled the end of Comanche hegemony over the Southern Plains," by Marini.

12. "The Rise and Fall of the Comanche Empire," *NPR,* 20 May 2011; *Empire of the Summer Moon,* by S.C. Gwynne (N.p.: Scribner, 2010).

13. "Last Days of the Comanches," *Texas Monthly* on line, n.d.

14. Holahan, David, review of *Empire of the Summer Moon.*

15. "Cheyenne War Parties and Battle Tactics," *Colorado History* on line, n.d.

16. "Quana Parker . . . ," *The Oklahoman,* 20 July 2002.

17. "An Indian Fight: Battle Between the Sioux and Chippewas," *St. Paul Pioneer* (MN), 12 July 1861.

18. Dr. Haha Lung, *Knights of Darkness: Secrets of the World's Deadliest Night Fighters* (Boulder, CO: Paladin Press, 1998), p. 21.

19. "The Rise and Fall of the Comanche Empire," *NPR,* 20 May 2011.

20. "Art of Manliness Podcast #526," interview of Sam Gwynne by Brett of 17 July 2017.

21. "The Rise and Fall of the Comanche Empire," *NPR,* 20 May 2011.
22. Lung, *Knights of Darkness.*
23. *Phantom Soldier,* by Poole, chapt. 14.
24. *The Tiger's Way,* by Poole, pp. 226, 227.
25. "Jääkärijoukkueen ja -Ryhmä Käsikirja," *Rifle-Light Infantry Platoon and Squad Handbook,* Finnish Army, Helsinki, 2018, HN878, trans. John Manter, p. 127.
26. "Honesty: The Importance of Truth In Native American Culture, IPL.org; "Setting the Record Straight About Native Languages: 'Ishmay Ohmay [in Blackfoot]'," Native-Languages.org;; "No Word for Lying [in Apache]," Language Log.
27. *The Wiles of War,* comp. and trans. Sun Haichen (Beijing: Foreign Languages Press, 1991), p. 328; *Phantom Soldier,* by Poole, p. 30; Stefan H. Verstappen, *The Thirty-Six Strategies of Ancient China* (San Francisco: China Books & Periodicals, 1999), p. 196; *Mao Tse-tung: An Anthology of His Writings,* ed. Fremantle, p. 69; Maitland and McInerney, *Vietnam Experience: A Contagion of War,* p. 97.
28. "Cochise and the Bascom Affair," by Jay W. Sharp, *DesertUSA.com.*
29. "From Makin to Bougainville: Marine Raiders in the Pacific War," by Maj. Jon T. Hoffman, *Marines in WWII Commemorative Series* (Washington, D.C.: Marine Corps Hist. Ctr., 1995), pp. 1-5; *Mao Tse-tung: An Anthology of His Writings,* ed. Fremantle, p. 69.
30. *Wikipedia Encyclopedia,* s.v. "Nez Perce War."
31. Alvin M. Josephy, Jr., *The Nez Perce and the Opening of the Northwest* (New Haven: Yale Univ. Press, 1965), pp. 632–633, in *Wikipedia Encyclopedia,* s.v. "Nez Perce."
32. "Nez Perce & U.S. Cavalry," USDA Forest Service, n.d.
33. "Nez Perce Fight Battle of Big Hole," *History.com,* n.d.; "Nez Perce & U.S. Cavalry," Forest Service.
34. *Wikipedia Encyclopedia,* s.v. "Nez Perce War."
35. "Nez Perce Fight Battle of Big Hole," *History.com,* n.d.
36. "The Battle at Clark Fork," USDA Forest Service, n.d.
37. *Frontier Skills,* by Davis, p. 89.
38. Frontiersmen in Blue, by Robert Utley (New York: Macmillan, 1967), pp. 7-9, in *Frontier Skills,* by Davis, pp. 101, 102.
39. *Frontier Skills,* by Davis, p. 104.
40. Ibid., p. 103.
41. "The Seminole Wars," Florida Dept. of State, n.d.
42. Ibid.
43. "Indians Don't Attack at Night," *Jeff Arnold's West.*
44. *Frontier Skills,* by Davis, p. 110.

45. Thomas Cleary, preface to *The Art of War,* by Sun Tzu (Boston: Shambhala Pubs., 1988), pp. vii, viii.
46. Dr. Masaaki. Hatsumi, *The Essence of Ninjutsu* (Chicago: Contemporary Books, 1988), p. 12.
47. Stephen K. Hayes, *Ninjutsu: The Art of the Invisible Warrior* (Chicago: Contemporary Books, 1984), p. 153.
48. Hatsumi, *The Grandmaster's Book of Ninja Training,* p. 2.
49. Ibid., p. 28.
50. Ibid., p. 8.
51. Ibid., p. 4.
52. Ibid., p. 83.
53. Ashida Kim, *The Invisible Ninja: Ancient Secrets of Surprise* (New York: Citadel Press, 1983), p. 1.
54. Ibid., p. 23.
55. Vo Nguyen Giap, as quoted in "The Tiger and the Elephant: Viet Minh Strategy and Tactics."
56. "Historical Study—Night Combat," *DA Pamphlet 20-236* (Washington, D.C.: Hdqts. Dept. of the Army, 1953), pp. 19-21.
57. *The Tiger's Way,* by Poole, p. 68.
58. Xiaobing Li, *China's Battle for Korea: The 1951 Spring Offensive* (Bloomington, IN: Indiana Univ. Press, 2014), pp. xxv-xxvi.
59. Giap, "Once Again We Will Win," as quoted in *The Military Art of People's War,* pp. 264, 265.
60. Edward J. Drea, "Nomonhan: Japanese—Soviet Tactical Combat, 1939," *Leavenworth Papers No. 2* (Ft. Leavenworth, KS: Combat Studies Inst., U.S. Army Cmd. and Gen. Staff College, 1981), p. 88.
61. Pope John Paul II, *Crossing the Threshold of Hope,* pp. 205, 206.
62. Sun Tzu, *The Art of War,* trans. and with intro. by Samuel B. Griffith, foreword by B.H. Liddell Hart (New York: Oxford Univ. Press, 1963), p. 39.
63. "Ulysses S. Grant's Overland Campaign: Six Bloody Weeks," by Christopher Klein, *History Channel,* updated 21 April 2020.
64. "Winning the West the Army in the Indian Wars, 1865-1890," *Army Historical Series,* Office of the Chief of Mil. Hist. U.S. Army, chapt. 14

Chapter 9: *The Spanish-American War*

1. "71st Regiment Infantry New York Volunteers Spanish-American War," New York Museum and Veterans Research Center.
2. "The 22nd U.S. Infantry in the Spanish-American War," Spanish American War website.

3. "Battle of San Juan Hill," by Raymond K. Bluhm, *Encyclopedia Britannica.*
4. Ibid.
5. Ibid.
6. "The Battle of San Juan Hill and El Caney," by SGM Chris Kalili, SGM Brian Pierce, and MSG Linda Williams,
probable Sgt.Maj. Academy thesis, Ft. Bliss, 2005.
7. "Black Jack in Cuba: General John J. Pershing's Service in the Spanish-American War," by Kevin Hymel, U.S. Army Ctr. of Mil. Hist.
8. "The World of 1898: The Spanish-American War," Library of Congress Researchers, n.d.
9. "The Rough Riders Storm San Juan Hill, 1898," *Eyewitness to History,* 2004.
10. Ibid.
11. Ibid.
12. Ibid.
13. Ibid.
14. Ibid.
15. "This Old Gun: Model 1893 Spanish Mauser," by Garry James, *American Rifleman,* 17 February 2002.
16. "Model 1893 Spanish Mauser," *Military Guns Australia,* n.d.; *Wikipedia Encyclopedia,* s.v. "Battle of San Juan Hill."
17. "The Spanish Modelo 1893 Mauser Rifle." *Shooting Times* on line, 29 March 2017.
18. Ibid.
19. *Wikipedia Encyclopedia,* s.v. "Battle of San Juan Hill."
20. "The Spanish 7x57mm Mauser," Spanish American War website.
21. *Gung Ho: The Corp's Most Progressive Tradition,* by H. John Poole (Emerald Isle, NC: Posterity Press, 2012), chapt. 12; *Phantom Soldier,* by Poole, chapt. 7.

Chapter 10: *World War I*

1. Erich Ludendorff, *Ludendorff's Own Story,* vol. 1 (New York, 1919), p. 316, from Timothy L. Lupfer, "The Dynamics of Doctrine: The Changes in German Tactical Doctrine during the First World War," *Leavenworth Papers No. 4* (Ft. Leavenworth, KS: Combat Studies Inst., U.S. Army Cmd. and Gen. Staff College, 1981) p. 11.
2. Graeme C. Wyne, *If Germany Attacks* (N.p., n.d.; reprint Westport, CT, 1976), p. 295, from Timothy L. Lupfer, "The Dynamics of Doctrine: The Changes in German Tactical Doctrine during the First World War," *Leavenworth Papers No. 4* (Ft. Leavenworth, KS: Combat Studies Inst., U.S. Army Cmd. and Gen. Staff College, 1981)
p. 27.

3. Timothy L. Lupfer, "The Dynamics of Doctrine: The Changes in German Tactical Doctrine during the First World War," *Leavenworth Papers No. 4* (Ft. Leavenworth, KS: Combat Studies Inst., U.S. Army Cmd. and Gen. Staff College, 1981), p. 15.

4. Ibid.

5. Ibid.

6. Bruce I. Gudmundsson, *Stormtroop Tactics—Innovation in the German Army 1914-1918* (New York: Praeger, 1989), p. 94.

7. Lupfer, "The Dynamics of Doctrine," *Leavenworth Papers No. 4,* p. 13.

8. Ibid., p. 12.

9. Gudmundsson, *Stormtroop Tactics,* p. 94.

10. Steven L. Canby, "Classic Light Infantry and New Technology," produced under contract MDA 903-81-C-0207 for the Defense Advanced Research Project Agency (Potomac, MD: C&L Associates, 1983), p. 51.

11. Lupfer, "The Dynamics of Doctrine," *Leavenworth Papers No. 4,* p. 42.

12. *British Official History 1918,* vol. 1 (N.p., n.d.), p. 166, from Lupfer, "The Dynamics of Doctrine," *Leavenworth Papers No. 4,* p. 50.

13. Lupfer, "The Dynamics of Doctrine," *Leavenworth Papers No. 4,* p. 44.

14. Ibid., p. 43.

15. Gudmundsson, *Stormtroop Tactics,* p. 143.

16. Ibid., pp. 49, 191.

17. Canby, "Classic Light Infantry and New Technology."

18. Ibid.

19. "World War I: New Weapons, Same Old (Bad) Tactics," *National Interest,* by Michael Peck, n.d.

20. "What did men do on the frontline in World War One," Bitesize Topics, *BBC* on line, n.d.

21. Canby, "Classic Light Infantry and New Technology."

22. "World War I: New Weapons, Same Old (Bad) Tactics," by Peck.

23. Canby, "Classic Light Infantry and New Technology."

24. "What Kind of Techniques Did the French Use to Fight in WWI," by Ralph Heibutzki, *Seattle Post-Intelligencer,* n.d.

25. Ibid.

26. Ibid.

27. Ibid.

28. "Getting American Soldiers Trained in the Realities of Trench Warfare," America Battle Monuments Commission, 1 Nov. 2017.

29. Ibid.

30. "How the U.S. Military Learned to Learn in World War I: Lessons from the American Expeditionary Forces," by Michael Shurk, Rand Corp., *WarontheRocks*.com, 22 August 2018.

31. Ibid.
32. Ibid.
33. *The Marines,* editor in chief Edwin Howard Simmons, ed. J. Robert Moskin (Quantico, VA: Marine Heritage Fdn., 1998), pp. 50-55.
34. Îbid.
35. Ibid.
36. "Belleau Wood: Marines' Shrine," by Linda D. Kozaryn, Armed Forces Press Service, *Leatherneck Magazine*, 9 August 1998, p. 14.
37. *Their* Time in Hell: The 4th Marine Brigade at Belleau Wood, by George B. Clark (Pike, NH: The Brass Hat, 1996), pp. 40-66.
38. *The Marines,* editor in chief Simmons, pp. 50-55.
39. Ibid.
40. "From Amiens to Armistice: The Hundred Days Offensive," Imperial War Museums, n.d.
41. Ibid.
42. *Encyclopedia Britannica,* s.v. "trench warfare."
43. "From Amiens to Armistice," Imperial War Museums.
44. Ibid.
45. Ibid.
46. Ibid.
47. Ibid.
48. *Encylopedia Britannica,* s.v. "Battle of St. Mihiel."
49. "From Amiens to Armistice," Imperial War Museums.
50. Ibid.
51. Ibid.
52. Ibid.
53. "This Day in History September 26" *History Channel*, n.d.
54. Ibid.
55. Ibid.
56. Ibid.
57. "This Day in History September 29," *History Channel,* n.d.
58. "From Amiens to Armistice," Imperial War Museums.
59. Ibid.
60. "This Day in History September 29," *History Channel,* n.d.
61. *Encyclopedia Britannica,* s.v. "Battles of the Meuse-Argonne."
62. Gudmundsson, *Stormtroop Tactics,* p. 94.
63. *Okinawan Reckoning,* by Poole, chapts. 1-4.
64. "This Day in History, September 26, 1918: The Meuse-Argonne Campaign Begins," America Battle Monuments Commission, n.d.
65. "Meuse-Argonne," Library of Congress, n.d.
66. Caption on "Hell Fighters from Harlem," by H. Charles McBarron, Jr., *Wikimedia Commons.*
67. Memorandum for the record by H.J. Poole, after participating in several sweeps during the 15 months of infantry combat in Vietnam.

68. "Scotland—Rick Steve's Europe," *PBS's NC Explorer Channel,* July 2022.

69. Memorandum for the record by H.J. Poole, after getting within 20 yards of an enemy machinegun in Vietnam and then having to scurry back into a shell hole while it tried to kill him.

Chapter 11: *World War II*

1. "Role of the War Department.in WWII," National Archives, as retrieved in May 2022 from the following url: https://catalog.archives.gov/id/10450628; "The War Department and the Army Records," U.S. National Archives and Records Service, as retrieved in June 2022 from the following url: https://www.archives.gov/research/holocaust/finding-aid/military/rg-107.html.

2. *Encyclopedia Britannica,* s.v. "The Joint Chiefs of Staff."

3. "History of the Unified Command Plan 1946–2012," by Edward J. Drea, Ronald H. Cole, Walter S. Poole, James F. Schnabel, Robert Watson, and Willard Webb (N.p.: Hist. Office of the JCS, 2013).

4. "Army Chain of Command," LSU Army ROTC, as retrieved in June 2022 from the following url: https://www.lsu.edu/hss/milsci/cadre/army_chain_of_command.php; *Wikipedia Encyclopedia,* s.v. "Organizational Structure of the United States Department of Defense."

5. *Gung Ho,* by Poole, chapts. 1, 2, and 6.

6. Memorandum for the record, by H.J. Poole, after finding such a paper early in his writing career.

7. "From Makin to Bougainville," by Hoffman, pp. 1-5.

8. *Six Strategies for War: The Practice of Effective Leadership,* trans. Alan Chong (Singapore: Asiapac Books, 1993), pp. 91, 86.

9. *Mao Tse-tung: An Anthology of His Writings,* ed. Fremantle, p. 69; Stefan H. Verstappen, *The Thirty-Six Strategies of Ancient China* (San Francisco: China Books & Periodicals, 1999), p. 196; *The Wiles of War,* comp. and trans. Sun Haichen, p. 328.

10. Lt.Col. Whitman S. Bartley, *Iwo Jima: Amphibious Epic*, Washington, D.C.: Hist. Branch, HQMC, 1954, pp. 128, 129; "Into the Meatgrinder," chapt 6, of *Battle for Iwo Jima,* by Robert Leckie (N.p.: I Books, 2004); *The Iwo Alamo,* by Poole, chapt. 9.

11. *Peleliu Progress,* by Poole, chapt. 9.

12. *The First Marine Division on Okinawa 1 April - 30 June 1945,* by James R. Stockman (N.p.: Historical Div., HQMC, n.d.), pp. 40, 41.

13. *Sands of Iwo Jima,* video (Hollywood: Republic Pictures, 1980.

14. *Gung Ho,* by Poole, pp. 74-79.

15. *Marines in the Central Solomons,* by Maj. John N. Rentz, USMC Historical Monograph (Washington, D.C.: Hist. Branch, HQMC, 1952), chapt. 4, pp. 113, 114.

16. "From Makin to Bougainville," by Hoffman, p. 33.

17. Yahara, *The Battle for Okinawa,* pp. 63, 64.

18. Huber, "Japan's Battle for Okinawa, April - June 1945," *Leavenworth Papers No. 18,* chapt. 3.

19. *Saving Private Ryan,* DVD, Paramount Pictures, n.d.

20. *Okinawa: The Last Battle,* by Appleman et al, p. 214.

21. "How a German Squad Attacks a Position (WWII)," *Footage Archive,* from a training film entitled "The German Infantry Squad in Action: A Demonstration of Minor Field Tactics."

22. "The Evolution of the U.S. Army Infantry Squad: Where Do We Go from Here: Determining the Optimum Infantry Squad Organization for the Future," by Maj. S.E. Hughes, *monograph* (Ft. Leavenworth, KS: School of Advanced Mil. Studies, U.S. Army Cmd. and Gen. Staff College, 1994)

23. The Infantry Conference, Report of Committee "B" on Tactics and Techniques (Ft. Benning, GA: The U.S. Army Infantry School, 1946), pp. 5-6, in "The Evolution of the U.S. Army Infantry Squad," by Hughes, p. 6.

24. Robert Dupree and Horace Homesly, Jr., " History of United States Army Squads and Platoons: 1935-1967" (Ft. Benning, GA: Combat Developments Cmd. Infantry Agency, 1967), p. 37, in "The Evolution of the U.S. Army Infantry Squad," by Hughes, p. 7.

25. "Night Attack – The Forsaken Tactic," *Craigchapman.com,* 31 October 2017.

26. *Iwo Alamo,* by Poole, p. 212.

27. "Battle of Peleliu" *Facts and Details,* n.d., as retrieved in December 2020 from the following url: http://factsanddetails.com/asian/ca67/sub428/entry-5336.html.

28. "Map May Hold Key," by David. S. Cloud, *Los Angeles Times* on line, 3 May 2015.

29. *Peleliu Progress,* by Poole, p. 216.

30. "Place of the Loyal Samurai," by Daniel Weiss, *Archaeology Magazine*, July/August 2019.

31. "Iwo Jima and Okinawa: Death at Japan's Doorstep," *National WWII Museum* website, 10 July 2017.

32. Bill D. Ross, *Iwo Jima: Legacy of Valor (*New York: Vintage, 1986), pp. 293, 294.

33. Bartley, *Iwo Jima: Amphibious Epic,* pp. 108, 109.

34. Yahara, *The Battle for Okinawa,* pp. xii, xiii.

35. *Okinawa: The Last Battle,* by Appleman et al, p. 194.

36. Bruce Cumings, in "The Battle of the Minds" segment of *Korea—the Unknown War* (London: Thames TV in assoc. with WGBH Boston, 1990), NC Public TV, n.d.

37. Member of U.S. Strategic Bombing Survey Team, in "Last Days of World War II," *History Channel,* 28 March 1999.

38. Col. Joseph H. Alexander, "Closing In: Marines in the Seizure of Iwo Jima," *Marines in World War II Commemorative Series* (Washington, D.C.: Hist. and Museums Div., HQMC, 1994), p. 13; Bartley, *Iwo Jima: Amphibious Epic,* p. 50; *The Spearhead* (5th Marine Division's unofficial website), s.v. "Battle of Iwo Jima."

39. "The Battle of Biak: 'A Terrifying Glimpse Into the Soul of Mankind," by David Alan Johnson, *Warfare History Network,* 30 December 2018.

40. *Phantom Soldier*, by Poole, chapt. 4.

41. Ibid., p. 125.

42 *Okinawan Reckoning,* by Poole, p. 230.

43. "Ryukyus: The U.S. Army Campaigns of WWII," *CMH Pub. 72-35,* by Arnold G. Fisch, Jr. (N.p.: U.S. Army Ctr. of Mil. Hist., 10 October 2012), p. 31.

44. "In Normandy, American Ingenuity Defeated the Impenetrable," by Herb Ditzel III, *The New York Times* on line, 5 June 1994.

45. Trevor N. Dupuy, David L. Bongard, and Richard C. Anderson Jr., *Hitler's Last Gamble* (New York: Harper Collins, 1994), pp. 333, 341, 342, 463-477.

46. "The Last Days of World War II," *History Channel,* 28 March 1999.

47. Charles B. MacDonald, *A Time for Trumpets* (New York: William Morrow, 1985), pp. 604, 605.

48. Dupuy et al, *Hitler's Last Gamble,* pp. 498, 499.

49. Allan R. Millett and Peter Maslowski, *For the Common Defense* (New York: Macmillan, 1984), p. 453.

50. *Gung Ho,* by Poole, chapts. 9, 10.

Chapter 12: *The Korean War*

1. "Memories of Massacres Were Long Suppressed Here: Tourists Now Retrace the Atrocities," by Choe Sang-Hun ("South Korea Dispatch"), *New York Times* on line, Asia Pacific Edition, 28 May 2019.

2. *Afrique: A Warning for America.* By H. John Poole (Emerald Isle, NC: Posterity Press, 2015), chapt. 4.

3. "Memories of Massacres Were Long Suppressed Here," by Choe Sang-Hun.

4. "An Arrogant Display of Strength" segment of *Korea—the Unknown War* (London: Thames TV in assoc. with WGBH Boston, 1990), NC Public TV.

5. Chung Il-Wan, S. Korean Constabulary, in "An Arrogant Display of Strength" segment of *Korea—the Unknown War*, NC Public TV.

6. Bert Hardy, photographer for the *Picture Post,* in "There is No Substitute for Victory" segment of *Korea—the Unknown War* (London: Thames TV in assoc. with WGBH Boston, 1990), NC Public TV.

7. "The Korean War Atrocities No One Wants to Talk About," by Jim Bovard, *The American Conservative,* 26 June 2020.

8. Rt.Hon. George Younger M.P., future British Defense Minister (then member of Argyll & Southern Highlanders), in "An Entirely New War" segment of *Korea—the Unknown War,* NC Public TV.

9. Col. Joseph H. Alexander, "Battle of the Barricades: U.S. Marines in the Recapture of Seoul," *Marines in the Korean War Commemorative Series* (Washington, D.C.: Hist. and Museums Div., HQMC, 2000), pp. 8-10.

10. *The Korean War: History and Tactics,* ed. by David Rees (New York: Crescent Books, 1984), p. 45.

11. Alexander, *Battle of the Barricades,* pp. 8-10.

12. Ibid.

13. "An Entirely New War" segment of *Korea—the Unknown War,* NC Public TV.

14. Max Hastings, *The Korean War* (New York: Touchstone, Simon & Schuster, 1987), p. 196.

15. "An Entirely New War," *Korea—the Unknown War,* NC Public TV.

16. Maj.Gen. Edwin Simmons USMC (Ret.), in "An Entirely New War" segment of *Korea—the Unknown War,* NC Public TV.

17. *Encyclopedia Britannica,* s.v. "U.S. Department of Defense," as retrieved in July 2022 from the following url: https://www.britannica.com/topic/US-Department-of-Defense.

18. Ibid., s.v. "The Joint Chiefs of Staff."

19. Chesty Puller as quoted in *Marine,* by Burke Davis (New York: Bantam, 1964), p. 390.

20. Ibid., p. 325.

21. Richard E. Ecker, *Battles of the Korean War: A Chronology* (Jefferson, NC: McFarland & Company, 2004), p. 32; Roy Appleman, *Escaping the Trap: The US Army X Corps in Northeast Korea,* 1950 (College Station, TX: Texas A&M Univ., 1990)."

22. *Mao's Generals Remember Korea,* trans. and ed. Xiaobing Li, Allan R. Millett, and Bin Yu (Lawrence, KS: Univ. Press of Kansas, 2001), pp. 153-155.

23. *This Kind of War,* by T.R. Fehrenbach (Washington, D.C.: Brassey's, 1994), originally published in New York by MacMillan in 1963), pp. 346-348.

24. Poole, *Phantom Soldier*, chapt. 6 and fig. 6-8.

25. *Mao's Generals Remember Korea,* trans. and ed. Xiaobing Li, Millett, and Bin Yu, p. 154.

26. Statistics from the Memorial Museum of the WRUSAK (War to Resist the U.S. and Aid Korea) at Dandong, China; and Tan, *Kangmei yuanchao zhanzheng* [The WRUSAK], 229; both in *China's Battle for Korea: The 1951 Spring Offensive, by* Xiaobing Li, pp. 118, 219.

27. *This Kind of War,* by Fehrenbach, p. 355.

28. Ibid., p. 348.

29. Ibid.

30. *Mao's Generals Remember Korea,* trans. and ed. Xiaobing Li, Millett, and Bin Yu, p. 154.

31. Du, *Zai zhiyuanjun zongbu* [at the CPVF General HQ], 264; Tan Jingjiao, *Kangmei yuanchao zhanzheng* [The WRUSAK] (Beijing: Zhongguo shehui kexue chubanshe, 1990), 175; Li Qingshan, *Zhiyuanjun yuanchou jishi* [the CPVF records of aiding Korea] (Beijing: Zhonggong dangshi chubanshe, 2008), 340. 351-352; all in *China's Battle for Korea: The 1951 Spring Offensive, by* Xiaobing Li, pp. 217, 218.

32. Joshua W. Montandon, *Battle for the Punchbowl: The U. S. First Marine Division's 1951 Fall Offensive of the Korean War,* Masters Thesis for Univ. of North Texas, 2007, pp. 133-140.

33. *This Kind of War,* by Fehrenbach, p. 353.

34. Ibid., p. 359.

35. *Wikipedia Encyclopedia,* s.v. "The Battle of Kumsong."

36. Cumings, "The Battle of the Minds," *Korea—the Unknown War,* NC Public TV.

37. "The Battle of the Minds," *Korea—the Unknown War,* NC Public TV.

38. Yahara, *The Battle for Okinawa,* pp. xii, xiii.

39. Ibid., p. 92.

40. *Phantom Soldier,* by Poole, chapt. 7.

41. "The Evolution of the U.S. Army Infantry Squad," by Hughes, in "The Infantry Squad Part 2: How Did We Get Here," by Chris Raynor. *NCO Journal,* 23 March 2018.

42. "Development of the Squad: Historical Analysis," by Ahmed Hashim, with contributions from Lt.Gen. Paul Van Riper USMC (Ret.) (Alexandria, VA: Ctr. for Naval Analyses, 2000), pp. 26, 27.

43. "Development of the Squad: Historical Analysis," by Hashim, p. 23

44. J.C. Fry, "Battle Drill," *Combat Forces Journal,* April 1953, pp. 18-22.

45. *Phantom Soldier,* by Poole, pp. 22, 29.

46. "An Entirely New War" segment of *Korea—the Unknown War,* NC Public TV.

47. *Phantom Soldier,* by Poole, chapt. 7.

Chapter 13: *The Vietnam War*

1. "Development of the Squad: Historical Analysis," by Hashim, p. 36.
2. Col. David H. Hackworth U.S. Army (Ret.) and Julie Sherman, *About Face* (New York: Simon & Schuster, 1989), p. 594.
3. Al Hemingway, *Our War Was Different* (Annapolis, MD: Naval Inst. Press, 1994), pp. 3 and 4, in *Wikipedia Encyclopedia,* s.v. "Combined Action Platoon."
4. Unofficial website of the CAP Program, as retrieved from this url: http://capmarine.com/index.htm; *Wikipedia Encyclopedia,* s.v. "Lewis William Walt" and "Victor H. Krulak."
5. Unofficial website of the CAP Program.
6. Ibid.; *Wikipedia Encyclopedia,* s.v. "Combined Action Platoon."
7. Richard D. Jackson, *Yesterdays Are Forever: A Rite of Passage through the Marine Corps and Vietnam* (Galt, CA: Working Title Publishing, 2007), p. 113.
8. III MAF Staff Letter, as quoted in *Wikipedia Encyclopedia,* s.v. "Combined Action Platoon"; Hemingway, *Our War Was Different,* p. 9.
9. "Westmoreland." *History Channel* on line, 29 May 2019.
10. Ibid.
11. Ibid.
12. "Lyndon B, Johnson: Foreign Affairs," by Kent Germany, UVA Miller Center, n.d.
13. *Wikipedia Encyclopedia,* s.v. "Vietnamization."
14. "Fact Sheet on the Combined Action Force," III MAF, 31 March 1970.
15. "List of Killed in Action: Combined Action Program, 1965-1971," as compiled and published by HQMC, and retrieved from Assoc. of the Combined Action Program website.
16. "The Future Role of the Combined Action Program," by Major Michael Duane Weltsch USMC, thesis for U.S. Army's Cmd. and Gen. Staff College, 1991, p. 104; Hemingway, *Our War Was Different,* p. 11.
17. "Malaysia: A Country Study," DA Pamphlet 550-45, *Area Handbook Series* (Washington, D.C.: Hdqts. Dept. of the Army, 1984), pp. 50-59.
18. Memo for the record from H.J. Poole, during his first tour in Vietnam.
19. Memo for the record from H.J. Poole, during his second tour in Vietnam.

Chapter 14: *The War in Iraq*

1. *Wikipedia Encyclopedia,* s.v. "Joint Chiefs of Staff."

2. "About the Joint Chiefs of Staff," www.jcs.mil, August 2022.
3. Scott Peterson, "Marine, Insurgent Tactics Evolve," *Christian Science Monitor,* 17 November 2004, p. 6.
4. Louis Meixler (AP). "Extremists Send Recruits into Iraq via Iranian Border," *Jacksonville Daily News* (NC), 8 November 2004, p. 4A.
5. Tini Tran (AP), "Marines, Iraqis Launch Offensive," *Jacksonville Daily News* (NC), 24 November 2004, pp. 1A, 2A.
6. "Challenges to Iran's Role in Iraq in the Post-Soleimani Era," by Hamidreza Azizi, *Stiftung Wissenschaft und Politik,* 22 July 2021.
7. *Tactics of the Crescent Moon,* by Poole, chapt. 2.
8. "Syria's Civil War Produces a Clear Winner: Hezbollah," by Maria Abi-Habib, *Wall Street Journal* on line, 3 April 2017; *Wikipedia Encyclopedia,* s.v. "Iranian involvement in the Syrian civil war."
9. "Yemen's Houthis and the Expansion of Iran's Axis of Resistance," by Katherine Zimmerman, American Enterprise Inst., March 2022.
10. "Iran Country Study," DA PAM 550-68, *Area Handbook Series* (Washington, D.C.: Hdqts. Dept. of the Army, 1989), p. 267.
11. Sepehr Zabih, *The Iranian Military in Revolution and War* (London: Routledge, 1988), p. 14.
12. Ibid., pp. 210-212.
13. Ibid., pp. 14, 211.
13. Ibid., p. 143.
14. Ibid., pp. 152, 160.
16. Robin Wright, "A Reporter at Large—Tehran Summer," *The New Yorker,* 5 September 1988, p. 40, as quoted in *Warriors of Islam: Iran's Revolutionary Guard,* by Katzman, (Boulder, CO: Westview Press, 1993) p. 63.
17. Kenneth Katzman, *Warriors of Islam: Iran's Revolutionary Guard* (Boulder, CO: Westview Press, 1993), pp. 82-84.
18. Roderick MacFarquhar and Michael Schoenhals, *Mao's Last Revolution* (Cambridge, MA: Harvard Univ. Press, 2006), pp. 125, 126; "What Was the Cultural Revolution," by Kallie Szczepanski, *About.com,* n.d.
19. Katzman, *Warriors of Islam: Iran's Revolutionary Guard,* pp. 84, 85.
20. GlobalSecurity.org, s.v. "[DPRK] Reconnaissance Bureau."
21. "North Korea, a Country Study," *DA Pamphlet 550-81,* 4th ed., ed. Savada, 1993, pp. 258-261.
22. Zabih, *The Iranian Military in Revolution and War,* p. 14.
23. Ibid., p. 160; Katzman, *Warriors of Islam: Iran's Revolutionary Guard,* pp. 82-84.
24. *Afrique,* by Poole, chapt. 4.
25. *Wikipedia Encyclopedia,* s.v. "Iranian involvement in the Syrian civil war."

26. "Syria's Civil War Produces a Clear Winner: Hezbollah," by Maria Abi-Habib, *Wall Street Journal* on line, 3 April 2017.

27. *Encyclopedia Britannica,* s.v. "Quds Force."

28. Ibid., s.v. "suicide bombing."

29. Sergeant of South African descent from U.S. Marine tank battalion scout platoon, in e-mail to author on 20 May 2017.

30. *Encyclopedia Britannica,* s.v. "the Surge."

Chapter 15: *The War in Afghanistan*

1. Ali A. Jalali. and Lester W. Grau. *Afghan Guerrilla Warfare: In the Words of the Mujahideen Fighters* (St. Paul, MN: MBI Publishing, 2001), originally published as *The Other Side of the Mountain.* Quantico, VA: Marine Corps Combat Development Cmd., 1995), p. xvi.

2. Robert D. Kaplan, *Soldiers of God: With Islamic Warriors in Afghanistan and Pakistan,* revised edition (New York: Vintage Books, 2001), p. 17.

3. Ibid., p. 235.

4. Brigadier Mohammad Yousaf, *The Silent Soldier: The Man Behind the Afghan Jehad* (South Yorkshire, UK: Leo Cooper, n.d.).

5. Brigadier Mohammad Yousaf, and Maj. Mark Adkin. *Bear Trap: Afghanistan's Untold Story* (South Yorkshire, UK: Leo Cooper, n.d.).

6. "Backgrounder on Afghanistan: History of the War," as extracted from www.geocities.com on 29 July 2004.

7. Jason Burke, "Waiting for a Last Battle with the Taliban," *The Observer* (UK), 27 June 1999.

8. "Return of the Taliban," *PBS's Frontline,* NC Public TV, 8 January 2007

9. "Pakistan's Involvement in Narco-Terrorism," by John Ward Anderson and Kamran Khan, *Congressional Record,* 3 October 1994, as retrieved in August 2022 from the following url: https://irp.fas.org/congress/1994_cr/h941003-terror-pak.htm.

10. Jalali and Grau, *Afghan Guerrilla Warfare,* editor's commentary, p. 72.

11. Tom A. Peter, "U.S. Soldiers: Iraq, This Isn't," *Christian Science Monitor,* 29 November 2009, p. 10.

12. Yousaf and Adkin, *Bear Trap.*

13. Ibid.

14. Ansnullah Wazir (AP), "Fight Wanes between Troops, Suspected al-Qaida," *Jacksonville Daily News* (NC), 22 March 2004, p. 4A.

15. "Return of the Taliban."

16. S.Sgt. Chris Ronin USMC (Ret.) [former Marine scout/sniper and special operator], in conversations with author from 2017 to 2019.

17. *2020 Drug Threat Assessment,* DEA website, 2 March 2021.

18. *Afrique,* by Poole, chapt. 14.

19. Multi-tour U.S. Army Special Forces veteran of wars in Iraq and Afghanistan and expert mantracker, in e-mail and telephone conversations with author between June 2013 and May 2015.

20. Ibid.

21. *Expeditionary Eagles: Outmaneuvering the Taliban,* by H. John Poole (Emerald Isle, NC: Posterity Press, 2010), chapt. 15.

22. "This Day in History, October 7," *History Channel,* n.d., as retrieved in August 2022 from the following url: https://www.history.com/this-day-in-history/u-s-led-attack-on-afghanistan-begins.

23. Ann Scott Tyson, "Going in Small in Afghanistan," *Christian Science Monitor,* 14 January 2004, pp. 11-13.

24. Ibid.

25. Tim McGirk, "Battle in 'the Evilest Place'," *Time Magazine,* 3 November 2003, East Asia ed., pp. 32-35.

26. Patrick McDowell (AP), "Pakistan Goes After Terrorists on Border," *Jacksonville Daily News* (NC), 9 January 2004, p. 6A.

27. *ABC's Morning News,* 8 May 2005.

28. Yousaf and Adkin, *Bear Trap.*

29. Robert Burns (AP), "Afghan Progress Is Cited," *Jacksonville Daily News* (NC), 22 May 2005, pp. 1A, 7A.

30. Daniel Cooney (AP), "U.S. Kills 12 Insurgents in Afghanistan," *Jacksonville Daily News* (NC), 23 May 2005, p. 4A.

31. "Two Insurgents Killed in Afghanistan Fight," World Briefs Wire Reports (AP), *Jacksonville Daily News* (NC), 25 May 2005, p. 10A.

32. Daniel Cooney (AP), "Rocket Attacks Kills 2 Americans at Base in Afghanistan," *Jacksonville Daily News* (NC), 9 June 2005, p. 4A.

33. "U.S. Soldier, 7 Militants Die in Afghan Firefight," World Briefs Wire Reports (AP), *Jacksonville Daily News* (NC), 11 June 2005, p. 5A.

34. Cooney, "Rocket Attacks Kills 2 Americans at Base in Afghanistan," p. 4A.

35. Noor Khan (AP), "Four American Soldiers Hurt in Afghanistan Blast," *Jacksonville Daily News* (NC), 14 June 2005, p. 4A.

36. Pakistani TV News, 31 May 2005.

37. Noor Khan (AP), "Afghan Cops Abducted by Taliban," *Jacksonville Daily News* (NC), 19 June 2004, p. 4A.

38. "76 Militants Die in Afghan Battles," World Brief Wire Reports (AP), *Jacksonville Daily News* (NC), 23 June 2004. p. 5A.

39. Daniel Cooney (AP), "Violence Spikes in Afghanistan," *Jacksonville Daily News* (NC), 26 July 2005, p. 6A.

40. "Suicide Bomber Kills Peacekeeper, Civilian in Afghan Capital," World Brief Wire Reports (AP), *Jacksonville Daily News* (NC), 28 January 2004.

41. Pauline Jelinek (AP), "U.S. Plans Spring Offensive in Afghanistan," *AOL News,* 30 January 2004.

42. "NATO Force Could Patrol Afghanistan," from AP, *Jacksonville Daily News* (NC), 5 August 2005, p. 4A; *ABC's Morning News,* 18 August 2009; *ABC's Nightly News,* 21 August 2005.

43. Jason Straziuso, Jason (AP), "Afghan Bomber Deadlier," *Jacksonville Daily News* (NC), 3 June 2007, pp. 1A, 8A; Richard Clarke, on *ABC's Morning News,* 7 June 2007; *ABC's Nightly News,* 6 June 2007; "Military: Iran Teaching Iraqi Militia to Build Bombs," from AP, *Jacksonville Daily News* (NC), 12 April 2007, p. 5A; Scott Peterson, "Shiites Rising—Part One," *Christian Science Monitor,* 6 June 2007, pp. 1, 13-16.

44. *Expeditionary Eagles,* by Poole, chapt. 10.

45. *ABC's Nightly News,* 31 August 2009.

46. "General Seeks New Afghan Approach," *Philadelphia Inquirer,* 1 August 2009.

47. President Barack Obama, in his address to the nation, on *ABC TV,* 1 December 2009.

48. *ABC's Nightly News,* 11 January 2010.

49. "The fall of Kabul: a 20-year mission collapses in a single day," by Emma Graham-Harrison and Luke Harding, *The Guardian* (UK) on line, 15 August 2021.

50. "The U.S. deal with the Taliban destroyed Afghans' military morale," by Bill Chappell, *NPR* on line, 18 May 2022.

51. Ibid.

52. Ibid.

53. Ibid.

54. *Peleliu Progress,* by Poole, chapt. 5.

Chapter 16: *Pentagon Has Failed to Evolve Tactically*

1. *Airland Battle Doctrine,* by Douglas W. Skinner (Alexandria, VA: Center for Naval Analyses, 1988), p. 29.

2. William S. Lind, *Maneuver Warfare Handbook* (Boulder, CO: Westview Press, 1985), p. 25.

3. *The Iwo Alamo,* by Poole, chapt. 13.

4. *Peleliu Progress,* by Poole, chapt. 9.

5. Huber, "Japan's Battle for Okinawa, April - June 1945," *Leavenworth Papers No. 18,* chapt. 3.

6. Gen. Mikhail Surkov, as quoted in "The World Turned Upside Down: Military Lessons of the Chechen War," by Anatol Lieven, *Armed Forces Journal International,* August 1998, pp. 40, 41.

7. *The Tiger's Way,* by Poole, chapt. 12.
8. *North Korea's Hidden Assets,* by Poole, chapts. 14-17.
9. Gen. Hoang Van Thai, *How South Vietnam Was Liberated* (Hanoi: The Gioi Publishers, 1996), p. 254.
10. Ibid., pp. 237, 238.
11. *Okinawan Reckoning,* by Poole, chapt. 9.
12. *You Are Worth It: Building a Life Worth Fighting For,* by Kyle Carpenter, Don Yaeger, et al (N.P.: HarperAudio. 2019).
13. Memorandum for the record by H.J. Poole, after conducting infantry training aboard many Army and Marine Bases until 2013.
14. Pope John Paul II, *Crossing the Threshold of Hope,* pp. 205, 206.

Chapter 17: *Shock and Awe Versus Deceptive Evasion*

1. Bartley, *Iwo Jima: Amphibious Epic,* pp. 108, 109.
2. "Warfighting," *MCDP 1,* pp. 42, 43.
3. Mao, as quoted in *China's Battle for Korea: The 1951 Spring Offensive, by* Xiaobing Li, pp. xxv-xxvi.
4. *The Strategic Advantage: Sun Zi and Western Approaches to War,* ed. Cao Shan (Beijing: New World Press, 1997), p. 6.
5. Giap, "Once Again We Will Win," as quoted in *The Military Art of People's War*, pp. 264, 265.
6. Drea, "Nomonhan: Japanese—Soviet Tactical Combat, 1939," *Leavenworth Papers No. 2*, p. 88.
7. "An Hoa Project," by Warren Stansbury, n.d.; Photograph caption of An Hoa Explosion, *Pinterest,* n.d.
8. *5th Marines Command Chronology for 23 February 1969,* in "An Hoa Project"; *3rd 155mm Gun Battery Command Chronology for 23 February 1969,* in "An Hoa Project."
9. Charles Soard (experienced infantry scout and frequent visitor to Vietnam), in telephone conversation with author on 28 March 2001.
10. "Biên Hòa Air Base Ammo Dump Explosion Sapper Attack," by Arnold John Houchin, originally written on 12 January 1972, then published by the Assoc. for USAF Vietnam War Veterans of Air/Security Police, through *Vietnam War Stories,* in 2005.
11. "Lost Battles of the Vietnam War."
12. "Attacks on MAG-16, Hill 22, and the Attempted Attack on the Da Nang Airfield," *3rd Marines in Vietnam* website, n.d.; Memorandum for the record by H.J. Poole, after seeing the article in *Stars and Stripes Magazine.*
13. "Lost Battles of the Vietnam War."
14. Memorandum for the record from H.J. Poole. while "base commandant" of 27th Marines' compound south of Da Nang.
15. *The Strategic Advantage,* ed. Cao Shan, pp. 21, 55.

16. Le Duan, *Ve chien tranh nhan dan Viet Nam* (Hanoi: Nat. Political Publishing House, 1993), pp. 295, 297, from Ho Khang, *The Tet Mau Than 1968 Event in South Vietnam* (Hanoi: The Gioi Publishers, 2001), p. 32.

17. *Okinawa: The Last Battle,* by Appleman, appendix C, table 2.

18. "Platoon," movie, written and directed by Oliver Stone, Hemdale Film Corp., 1986.

19. *Tactics of the Crescent Moon: Militant Muslim Combat Methods,* by H. John Poole (Emerald Isle, NC: Posterity Press, 2004), chapt. 3.

20. *Encyclopedia Britannica,* s.v. "suicide bombing."

21. Ibid., s.v "suicide bomber."

Chapter 18: *Some New Hardware Discourages Maneuver*

1. *Gung Ho,* by Poole, p. 311.

2. *The Last Hundred Yards: The NCO's Contribution to Warfare,* by H.J. Poole (Emerald Isle, NC: Posterity Press, 1997), chapt. 9.

3. "World War I: New Weapons, Same Old (Bad) Tactics," by Peck.

4. John Shaw and the editors of Time-Life Books, *Red Army Resurgent* (Chicago: Time-Life Books, 1979), pp. 163-165.

5. *Peleliu Progress,* by Poole, chapt. 8.

6. Lung, *Knights of Darkness,* pp. 53-56.

7. Hatsumi, *The Essence of Ninjutsu,* pp. 15, 37.

8. *Night Movements*, trans. and preface by C. Burnett (Tokyo: Imperial Japanese Army, 1913; reprint Port Townsend, WA: Loompanics Unlimited, n.d.), pp. 23-29.

9. Stephen K. Hayes, *Warrior Path of Togakure* (Santa Clarita, CA: Ohara Publications, 1983), p. 52.

10. Lung, *Knights of Darkness,* p. 59.

11. Ibid., p. 64.

12. *Night Movements,* p. 26.

13. *Okinawa: The Last Battle,* by Appleman et al, p, 38.

14. "Forward Looking Infrared Radar," *Fandom,* n.d.

15. Memorandum for the record from H.J. Poole. while on Operation Maui Peak in 1968 in Vietnam.

16. "Fighting on Guadalcanal," *FMFRP 12-110* (Washington, D.C.: U.S.A. War Office, 1942), p. 25.

17. "The Stryker brigade combat team (SBCT) infantry . . . ," *FM 3-21.94,* Appendix —Javelin Employment, from GlobalSecurity.org.

18. Memorandum for the record by H.J. Poole, after 15 months as an infantry platoon or company commander in Vietnam.

19. *The Tiger's Way,* by Poole, pp. 173-176.

20. "German Infantry Anti-Tank Tactics 1941/1942: Eastern Front Edition," *Military History Visualized;* "Handbook on Japanese Military Forces." *TM-E 30-480.* Washington, D.C.: U.S. War Department, 1944, reprint, Baton Rouge, LA: LSU Press, 1991, p. 117.

21. Memorandum for the record by H.J. Poole, after 15 months as an infantry platoon or company commander in Vietnam.

22. Ibid.; *Phantom Soldier,* by Poole, pp. 37, 44-46.

23. *ABC's Morning News,* 12 September 2022.

24. *Mao's Generals Remember Korea,* trans. and ed. Xiaobing Li, Millett, and Bin Yu, pp. 53-57, 136, 137.

25. Naomi Segal (Jewish Telegraphic Agency), "IDF Absolved of Blame in Deaths of Naval Commandos in Lebanon," *San Francisco Jewish Community Publication,* 31 October 1997.

26. "Drone War Vietnam," by David Axe (Annapolis, MD: U.S. Naval Inst. Press, 2021).

27. Lovingood, Col. Howard J. USMC (Ret.), through several conversations with the author around 2020; "Overstressing Safety Is Leaving Momentum Behind," by Poole, *Army Magazine,* August 2020.

28. S.L.A. Marshall, *Soldier's Load and the Mobility of a Nation* (Quantico, VA: Marine Corps Assoc., 1980), pp. 69, 70.

29. "Mail Call," *DEFY TV,* 18 October 2022.

30. Memorandum for the record by H.J. Poole, while with 2/5 from August 1968 to March 1969.

Chapter 19: *Squad Tactics Still Vital to Modern War*

1. *Phantom Soldier,* by Poole, chapt. 7.

2. Joseph S. Bermudez, Jr., *North Korean Special Forces* (Annapolis, MD: Naval Inst. Press, 1998), p. 171.

3. Lung, *Knights of Darkness,* p. 12.

4. "Confucian Beliefs," *Facts and Details,* as retrieved in September 2022 from this url: factsanddetails.com/china.

5. *World History Encyclopedia,* s.v. "Shinto."

6. "Confucianism and Shintoism—Similarities and Differences in Belief and Practice," by Nixie Adams, *Interfaith Now,* 16 Sept. 2021.

7. *Night Movements*, trans. and preface by Burnett, pp. 23-32.

8. *Militant Tricks: Battlefield Ruses of the Islamic Insurgent,* by H. John Poole (Emerald Isle, NC: Posterity Press, 2005), chapts. 3 and 6.

9. *The Tiger's Way,* by Poole, chapt. 23.

10. M.Gy.Sgt.Bob O'Bday, USMC (Ret.) (member of 1/9 in Vietnam and machinegunner at Khe Sanh), in telephone conversation with author, 14 October 2022.
11. *Okinawan Reckoning,* by Poole, chapt. 21.
12. Memorandum for the record by H.J. Poole, after 15 months as an infantry platoon or company commander in Vietnam.
13. Pres. John F. Kennedy, quote from his 1962 West Point graduation speech, in "Unconventional Warfare Fundamentals," at *The Irregular Warrior* website.
14. Lanning and Cragg, *Inside the VC and the NVA,* pp. 211, 212.

Chapter 20: *Washington's Perspective*

1. "Tactics: Rethinking the Basis of Infantry Close Combat," by Lt. Col. David Kilcullen, *Australian Infantry Journal,* April 2015.
2. *Okinawan Reckoning,* by Poole, chapt. 14; Vic Taylor, "Hotel Company-Day Three," *Marine Corps Gazette,* April 2004.
3. "Papal Encyclical Urges Capitalism to Shed Injustices," by Peter Steinfels. *New York Times* on line, 3 May 1991.

Chapter 21: *Battlefield Misconceptions*

1. "Warfighting." *FMFM 1,* p. 28
2. Lanning and Cragg, *Inside the VC and the NVA,* pp. 211, 212.
3. Yahara, *The Battle for Okinawa,* pp. xii, xiii; *Okinawa: The Last Battle,* by Appleman et al, p. 194.
4. Hackworth and Sherman, *About Face,* p. 594.
5. *ABC's Nightly News,* around 29 June 2022.
6. "Gunsmoke," *Greenville Channel 7.3 (WITN),* 23 July 2022.
7. "Gen. Erskine quotes," as retrieved in August 2022 from this url: https://loveexpands.com/author/graves-b-erskine/.

Chapter 22: *The American Rifleman's Experience*

1. Maj.Gen. David B. Barker, as overheard by the author while under his command either at 4th Marine Division or Camp Lejeune.
2. *Stay off the Skyline: The Sixth Marine Division on Okinawa - An Oral History,* by Laura Homan Lacy (Washington, D.C.: Potomac Books, 2007).
3. *With the Old Breed: At Peleliu and Okinawa,* by E.B. Sledge (New York: Presidio Press, a division of Random House, 2007), p. 225.

4. Ibid., p. 78.
5. Ibid., pp. 79, 80.
6. Gen. George S. Patton, as quoted in October 2022 at the following url: https://www.brainyquote.com/quotes/george_s_patton_105114.

Glossary

AEF	American Expeditionary Force	U.S. troops deployed to Europe during WWI
AP	Associated Press	U.S. news service
ARVN	Army of the Republic of Vietnam	South Vietnamese military units
BAR	Browning Automatic Rifle	Heavy repeating rifle used mostly during WWII
CAP	Combined Action Platoon	Unit comprised of three PF squads, each having one U.S. Marine fireteam
CIA	Central Intelligence Agency	U.S. spy outfit.
CO	Commanding Officer	Unit officer in charge
CP	Command Post	Unit headquarters
CPVA	Chinese People's Volunteer Army	Korean War entrant
DMZ	Demilitarized Zone	No-man's land.
E&E	Escape and Evasion	Avoiding capture.
EFP	Explosively Formed Penetrator	Armor-piercing IED
FMFPac	Fleet Marine Forces Pacific	Parent command of Marine divisions on Iwo Jima

4GW	4th Generation Warfare	Four-arena conflict: martial, political, economic, religious/ psychological
FPL	Final Protective Line	Interlocking MG fire
G-2	Headquarters staff section	Intelligence provider
GI	Government Issue	Colloquial term for low-ranking member of U.S. Armed Forces
GPS	Global Positioning System	Device that resects satellite signals to determine location.
GW	Guerrilla Warfare	UW by tiny insurgent groups
HIG	Hezb i-Islami Gulbuddin	Afghan rebel outfit
HMMWV	High Mobility Multipurpose Wheeled Vehicle	Modern-day jeep
HQMC	Headquarters Marine Corps	Top echelon of USMC
IED	Improvised Explosive Device	Remotely set off mine
IJA	Imperial Japanese Army	Japan's ground units
IR	Infrared	Electromagnetic radiation with a wavelength slightly greater than red
IRGC	Iranian Revolutionary Guard Corps	Iranian commandos
ISI	Inter-Services Intelligence	Pakistani spy agency
ISIS	Islamic State of Iraq and Syria	Self-proclaimed caliphate to which other radical factions pledge allegiance

JCS	Joint Chiefs of Staff	President's military advisors
KISS	Keep It Simple Stupid	Good advice for any leader.
LITGB	Light Infantry Training Guidance Bureau	North Korean defense agency.
LP	Listening Post	Nighttime OP.
MACV	Military Assistance Command Vietnam	Adviser central in Southeast Asia
MG	Machinegun	Automatic small arm
MLR	Main Line of Resistance	Linear formation of defensive positions.
MoH	Medal of Honor	Highest U.S. award for valor.
MW	Maneuver Warfare	Way of fighting where surprise takes place of firepower.
NATO	North Atlantic Treaty Organization	Western military alliance
NCO	Noncommissioned Officer	Junior enlisted military leader.
NKPA	North Korean People's Army	Pyongyang's ground forces
NPR	National Public Radio	U.S. educational radio network
NVA	North Vietnamese Army	Hanoi's ground forces
NVG	Night Vision Goggle	Device to enhance a soldier's nighttime viewing capability
OP	Observation Post.	Viewing station

PBS	Public Broadcasting System	U.S. educational television network
PF	Popular Force	Vietnam militiaman
PFC	Private First Class	Second lowest U.S. military rank
PLA	People's Liberation Army	Chinese defense force
POW	Prisoner of War	Captured combatant.
PRC	Peoples' Republic of China	Communist China
RCT	Regimental Combat Team	Reinforced infantry regiment.
RPG	Rocket Propelled Grenade	Ordnance launched by Communist bazooka-like weapon
S-3	Headquarters staff section	Operations provider
2GW	2nd Generation Warfare	Focus is on killing foe soldiers; same as traditional U.S. style (attrition warfare).
SMAW	Shoulder-Launched Multi-purpose Assault Weapon	Hand-held bunker busting missile launcher
SNCO	Staff Noncommissioned Officer	Senior enlisted military leader.
SOP	Standard Operating Procedure	Normal activity.
TAOR	Tactical Area of Responsibility	Zone of action.
IIIMAF	III Marine Amphibious Force	Marine headquarters in Vietnam
3GW	3rd Generation Warfare	Avoids foe defenders to more easily get at cmd. & logistic hubs.

U.N.	United Nations	Worldwide alliance of countries.
U.S.	United States	America.
USMC	U.S. Marine Corps	America's amphibious force in readiness.
UW	Unconventional Warfare	Attacking through infiltration and defending through ambushes; part of 3GW.
VC	Viet Cong	Enemy guerrillas in Vietnam
WWI	World War I	First global conflict.
WWII	World War II	Second global conflict
WWIII	World War III	Third global conflict

Bibliography

U.S. Government Publications, News Releases, and Videos

"Across the Reef: The Marine Assault of Tarawa." By Col. Joseph H. Alexander, USMC (Ret.). *Marines in WWII Commemorative Series.* Washington, D.C.: Marine Corps Historical Center, 1993. As retrieved in April 2022 from url: https://www.nps.gov/parkhistory/online_books/npswapa/extcontent/usmc/pcn-190-003120-00/sec8.htm.

Airland Battle Doctrine. By Douglas W. Skinner. Alexandria, VA: Center for Naval Analyses, 1988. As retrieved in August 2022 from the following url: https://apps.dtic.mil/sti/pdfs/ADA202888.pdf.

Alexander, Col. Joseph H. "Battle of the Barricades: U.S. Marines in the Recapture of Seoul." *Marines in the Korean War Commemorative Series.* Washington, D.C.: History and Museums Division, Headquarters, Marine Corps, 2000.

Alexander, Col. Joseph H. "Closing In: Marines in the Seizure of Iwo Jima." *Marines in World War II Commemorative Series.* Washington, D.C.: History and Museums Division, Headquarters Marine Corps, 1994. As retrieved in January 2020 from the following urls: https://ia902707.us.archive.org/20/items/ClosingIn/ClosingIn.pdf; https://www.nps.gov/parkhistory/online_books/npswapa/extcontent/usmc/pcn-190-003131-00/pcn-190-003131-00/index.htm.

Bartley, Lt.Col. Whitman S. *Iwo Jima: Amphibious Epic.* Washington, D.C.: History Branch, Headquarters Marine Corps, 1954.

Canby, Steven L. "Classic Light Infantry and New Technology." Produced under contract MDA 903-81-C-0207 for the Defense Advanced Research Project Agency. Potomac, MD: C&L Associates, 1983.

"Development of the Squad: Historical Analysis." By Ahmed Hashim. With contributions from Lt.Gen. Paul Van Riper USMC (Ret.). Alexandria, VA: Center for Naval Analyses, 2000. As retrieved in January 2019 from the following url: www.dtic.mil/dtic/tr/fulltext/u2/1014512.pdf.

Drea, Edward J. "Nomonhan: Japanese—Soviet Tactical Combat, 1939." *Leavenworth Papers No. 2.* Ft. Leavenworth, KS: Combat Studies Institute, U.S. Army Command and General Staff College, 1981.

Drug Enforcement Agency website. At this url: https://www.dea.gov.

"Fact Sheet on the Combined Action Force," III MAF, 31 March 1970.

"Fighting on Guadalcanal." *FMFRP 12-110.* Washington, D.C.: U.S.A. War Office, 1942.

The First Marine Division on Okinawa 1 April - 30 June 1945. By James R. Stockman. N.p.: Historical Division, Headquarters Marine Corps, n.d.

"From Makin to Bougainville: Marine Raiders in the Pacific War." By Maj. Jon T. Hoffman. *Marines in WWII Commemorative Series.* Washington, D.C.: Marine Corps History Center, 1995.

"The Future Role of the Combined Action Program." By Major Michael Duane Weltsch USMC. Thesis for U.S. Army's Command and General Staff College, 1991.

"Handbook on German Military Forces." *TM-E 30-451.* Washington, D.C.: U.S. War Department, 1945. Reprint Baton Rouge, LA: LSU Press, 1990.

"Handbook on Japanese Military Forces." *TM-E 30-480.* Washington, D.C.: U.S. War Department, 1944. Reprint, Baton Rouge, LA: Louisiana State Unversity Press, 1991.

"Historical Study—Night Combat," *DA Pamphlet 20-236.* Washington, D.C.: Headquarters Department of the Army, 1953.

"History of the Unified Command Plan 1946–2012." By Edward J. Drea, Ronald H. Cole, Walter S. Poole, James F. Schnabel, Robert J. Watson, and Willard J. Webb. N.p,: History Office of the Joint Chiefs of Staff, 2013. As retrieved in June from this url: https://www,jcs.mil/Portals/36/Documents/History/Institutional/Command-Plan.pdf.

Huber, Thomas M. "Japan's Battle for Okinawa, April - June 1945." *Leavenworth Papers No. 18.* Fort Leavenworth, KS: Combat Studies Institute, U.S. Army's Command and General Staff College, 1990.

"The Infantry Squad Part 2: How Did We Get Here?" By Chris Raynor. *NCO Journal,* 23 March 2018. As retrieved in January 2018 from the following url: https://www.armyupress.army.mil/Journals/NCO-Journal/Archives/2018/March/Infantry-Squad-Part-2/.

"Iran Country Study." DA PAM 550-68. *Area Handbook Series.* Washington, D.C.: Headquarters Department of the Army, 1989.

Lupfer, Timothy L. "The Dynamics of Doctrine: The Changes in German Tactical Doctrine During the First World War." *Leavenworth Papers No. 4.* Fort Leavenworth, KS: Combat Studies Institute, U.S. Army Command and General Staff College, 1981.

"Malaysia: A Country Study." DA Pamphlet 550-45. *Area Handbook Series*. Washington, D.C.: Headquarters Department of the Army, 1984.

"The Marine Rifle Squad." *MCRP 3-10A.4.* Quantico, VA: Headquarters Marine Corps, August 2020.

Marines in the Central Solomons. By Maj. John N. Rentz. USMC Historical Monograph. Washington, D.C.: Historical Branch, Headquarters Marine Corps, 1952.

"North Korea, a Country Study." *DA Pamphlet 550-81.* 4th edition. Area Handbook Series. Edited by Andrea Matles Savada. Washington, D.C.: Federal Research Division, Library of Congress, 1993. As retrieved in April 2018 from this url: https://babel.hathitrust.org/cgi/pt?id=umn.31951d002715018;view=1up;seq=297.

Okinawa: The Last Battle. By Roy E. Appleman, James M. Burns, Russell A. Gugeler, and John Stevens. *United States Army in World War II Series.* Washington, D.C.: U.S. Army's Center of Military History, 2000.

"Pakistan's Involvement in Narco-Terrorism." By John Ward Anderson and Kamran Khan, *Congressional Record,* 3 October 1994. As retrieved in August 2022 from the following url: https://irp.fas.org/congress/1994_cr/h941003-terror-pak.htm.

"Role of the War Department.in WWII." National Archives, n.d. As retrieved in May 2022 from the following url: https://catalog.archives.gov/id/10450628.

"Ryukyus: The U.S. Army Campaigns of WWII." *CMH Pub. 72-35.* By Arnold G. Fisch, Jr. N.p.: U.S. Army Center of Military History, 10 October 2012.

"The Stryker Brigade Combat Team (SBCT) Infantry . . . ." *FM 3-21.94.* Appendix —Javelin Employment. From GlobalSecurity.org.

"Tactics." *MCDP 1-3.* Washington, D.C,: Headquarters Marine Corps, 1997.

2020 Drug Threat Assessment. Drug Enforcement Administration, 2 March 2021. As retrieved in August 2022 from this url: https://www.dea.gov/sites/default/files/2021-02/DIR-008-21%202020%20National%20Drug%20Threat%20Assessment_WEB.pdf.

"The War Department and the Army Records." U.S. National Archives and Records Service. As retrieved in June 2022 from the following url: https://www.archives.gov/research/holocaust/finding-aid/military/rg-107.html.

"The War of the American Revolution." By Robert W. Coakley and Stetson Conn. *CMH_Pub_70-6-1.* Washington, D.C.: U.S. Army's Center of Military History, 2010.

"Warfighting." *FMFM 1.* Washington, D.C.: Headquarters Marine Corps, 1989.

"Warfighting." *MCDP 1.* Washington, D.C,: Headquarters Marine Corps, 1997.
"Winning the West the Army in the Indian Wars, 1865-1890." *Army Historical Series.* Office of the Chief of Military History U.S. Army. Chapter 14. As retrieved from the following url: https://history.army.mil/books/AMH/amh-14.htm.

Civilian Publications

Analytical Studies, Websites, and Personal Conversations

ABC News. As seen on TV Channel 12 TV. WCTI, New Bern, NC.
Afrique: A Warning for America. By H. John Poole. Emerald Isle, NC: Posterity Press, 2015.
"The American Revolution: General Francis Marion, the Swamp Fox, and Marine Corps Fighting Doctrine." By Wilson. Doctoral Research, U.S. Marine Corps Research Center, n.d.
Battle for Iwo Jima. By Robert Leckie. N.p.: I Books, 2004.
Bermudez, Joseph S., Jr. *North Korean Special Forces.* Annapolis, MD: Naval Institute Press, 1998.
CAP Program unofficial website. As accessed through this url: http://capmarine.com/index.htm.
China's Battle for Korea: The 1951 Spring Offensive. By Xiaobing Li. Bloomington, IN: Indiana University Press, 2014.
"The Civil War Siege of Jackson, Mississippi." By Jim Woodrick. *Civil War Series.* N.p.: History Press, 18 April 2016.
Davis, Burke. *Marine.* New York: Bantam, 1964.
Dictionary (electronic). From its website, www.dictionary.com.
"Drone War Vietnam." By David Axe. Annapolis, MD: U.S. Naval Instititute Press, 2021. As retrieved in September 2022 from the following url: https://www.usni.org/press/books/drone-war-vietnam .
Dupuy, Trevor N., and David L. Bongard and Richard C. Anderson Jr. *Hitler's Last Gamble.* New York: Harper Collins, 1994.
Ecker, Richard E. *Battles of the Korean War: A Chronology.* Jefferson, NC: McFarland & Company, 2004.
Empire of the Summer Moon. By S.C. Gwynne. N.p.: Scribner, 2010.
Encyclopedia Britannica. From its website, https://www.britannica.com.
Encyclopedia.com. From its website, https://www.encyclopedia.com,
Expeditionary Eagles: Outmaneuvering the Taliban. By H. John Poole. Emerald Isle, NC: Posterity Press, 2010.
Fox, Col. Wesley L. *Marine Rifleman.* Dulles, VA: Brassey's, 2002.

Frontier Skills: The Tactics and Weapons That Won the American West. By William C. Davis. Guilford, CT: Lyons Press, 2003.

Gale Encyclopedia of Native American Tribes. Edited by Sharon Malinowski and Anna Sheets. New York: Gale, 1998.

Giap, Gen. Vo Nguyen. "Once Again We Will Win." In *The Military Art of People's War.* Edited by Russel Stetler. New York: *Monthly Review Press,* 1970.

Global Security. From its website, https://www.globalsecurity.org.

Gung Ho: The Corps' Most Progressive Tradition. By H. John Poole. Emerald Isle, NC: Posterity Press, 2012.

Hackworth Col. David H. U.S. Army (Ret.) and Julie Sherman. *About Face.* New York: Simon & Schuster, 1989.

Hastings, Max. *The Korean War.* New York: Touchstone, Simon & Schuster, 1987.

Hatsumi, Dr. Masaaki. *The Essence of Ninjutsu.* Chicago: Contemporary Books, 1988.

Hatsumi, Dr. Masaaki. *The Grandmaster's Book of Ninja Training.* Translated by Chris W. P. Reynolds. Chicago: Contemporary Books, 1988.

Hayes, Stephen K. *Ninjutsu: The Art of the Invisible Warrior.* Chicago: Contemporary Books, 1984.

Hayes, Stephen K. *Warrior Path of Togakure.* Santa Clarita, CA: Ohara Publications, 1983.

The Iwo Alamo. By H. John Poole. Emerald Isle, NC: Posterity Press, 2020.

"Jääkärijoukkueen ja -Ryhmä Käsikirja." Rifle-Light Infantry Platoon and Squad Handbook. Finnish Army. Helsinki, 2018. HN878. Translated by John Manter.

Jackson, Richard D. *Yesterdays Are Forever: A Rite of Passage Through the Marine Corps and Vietnam.* Galt, CA: Working Title Publishing, 2007.

Jalali, Ali A. and Lester W. Grau. *Afghan Guerrilla Warfare: In the Words of the Mujahideen Fighters.* St. Paul, MN: MBI Publishing, 2001. Originally published as *The Other Side of the Mountain.* Quantico, VA: Marine Corps Combat Development Command, 1995.

Jaynes, Gregory and the editors of Time-Life Books, *The Civil War: The Killing Ground—Wilderness to Cold Harbor.* Alexandria, VA: Time-Life Books, 1986.

Kaplan, Robert D. *Soldiers of God: With Islamic Warriors in Afghanistan and Pakistan.* Revised edition. New York: Vintage Books, 2001.

Khang, Ho. *The Tet Mau Than 1968 Event in South Vietnam.* Hanoi: The Gioi Publishers, 2001.

Kim, Ashida. *The Invisible Ninja: Ancient Secrets of Surprise.* New York: Citadel Press, 1983.

Lanning, Michael Lee and Dan Cragg. *Inside the VC and the NVA: The Real Story of North Vietnam's Armed Forces.* New York: Ivy Books, 1992.

The Last Hundred Yards: The NCO's Contribution to Warfare. By H.J. Poole. Emerald Isle, NC: Posterity Press, 1997.

Lind, William S. *Maneuver Warfare Handbook.* Boulder, CO: Westview Press, 1985.

Lovingood, Col. Howard J. USMC (Ret.). Through several conversations with the author around 2020.

Lung, Dr. Haha. *Knights of Darkness: Secrets of the World's Deadliest Night Fighters.* Boulder, CO: Paladin Press, 1998. As retrieved in March 2022 from the following url: https://books.google.com/books?id=4xT2-3gHK_8C&pg=PA21&lpg=PA21&dq=apache+indian+night+fighters&source=bl&ots=mC2-9tyAnq&sig=ACfU3U195mmMsUgJSFKXJqDjdOHtRXxj-g&hl=en&sa=X&ved=2ahUKEwj6kdDn8rn2AhVEoXIEHbApDREQ6AF6BAhFEAM#v=onepage&q=apache%20indian%20night%20fighters&f=false.

MacDonald, Charles B. *A Time for Trumpets.* New York: William Morrow, 1985.

MacDonald, John. *Great Battles of the Civil War.* New York: Macmillan U.S.A. in association with Marshall Editions, Limited, 1992.

MacFarquhar, Roderick and Michael Schoenhals. *Mao's Last Revolution.* Cambridge, MA: Harvard University Press, 2006.

Maitland, Terrence and Peter McInerney. *Vietnam Experience: A Contagion of War.* Boston, MA: Boston Publishing, 1968.

Militant Tricks: Battlefield Ruses of the Islamic Insurgent. By H. John Poole. Emerald Isle, NC: Posterity Press, 2005.

Mao's Generals Remember Korea. Translated and edited by Xiaobing Li, Allan R. Millett, and Bin Yu. Lawrence, KS: University Press of Kansas, 2001.

Mao Tse-tung: An Anthology of His Writing. Edited by Anne Fremantle. New York: Mentor, 1962.

The Marines. Editor in chief, Edwin Howard Simmons. Edited by J. Robert Moskin. Quantico, VA: Marine Heritage Foundation, 1998.

Marshall, S.L.A. *Soldier's Load and the Mobility of a Nation.* Quantico, VA: Marine Corps Association, 1980.

McPherson, James M. editor, and Richard Gottlief, managing editor. *Battle Chronicles of the Civil War 1864.* New York: Grey Castle Press, 1989.

The Military Art of People's War. Edited by Russel Stetler. New York: Monthly Review Press, 1970.

Montandon, Joshua W. *Battle for the Punchbowl: The U. S. First Marine Division's 1951 Fall Offensive of the Korean War.* Masters thesis for University of North Texas, 2007.

Multi-tour U.S. Army Special Forces veteran of wars in Iraq and Afghanistan and expert mantracker. In e-mail and telephone conversations with author between June 2013 and May 2015.

"Native Americans in the Revolutionary War." *History of Massachusetts Blog*. As as retreived in April 2022 from the following url: https://historyofmassachusetts.org/native-americans-revolutionary-war/.

"NATO Force Could Patrol Afghanistan." From Associated Press. *Jacksonville Daily News* (NC), 5 August 2005.

New World Encyclopedia. From its website, newworldencyclopedia.org.

Night Movements. Translated and preface by C. Burnett. Tokyo: Imperial Japanese Army, 1913. Reprint at Port Townsend, WA: Loompanics Unlimited, n.d.

Nolan, Keith William. *Operation Buffalo: USMC Fight for the DMZ.* New York: Dell Publishing, 1991.

North Korea's Hidden Assets. By H. John Poole. Emerald Isle, NC: Posterity Press, 2018.

Obama, President Barak. In his address to the nation. On *ABC Television,* 1 December 2009.

O'Bday, M.Gy.Sgt.Bob USMC (Ret.) (member of 1/9 in Vietnam and machinegunner at Khe Sanh). In telephone conversation with author, 14 October 2022.

Okinawan Reckoning. By H. John Poole. Emerald Isle, NC: Posterity Press, 2022.

Patton, Gen. George S. As quoted in October 2022 at the following url: https://www.brainyquote.com/quotes/george_s_patton_105114.

Peleliu Progress. By H. John Poole. Emerald Isle, NC: Posterity Press, 2021.

Phantom Soldier: The Enemy's Answer to U.S. Firepower. By H. John Poole. Emerald Isle, NC: Posterity Press, 2001.

Robert Rogers' Rules for the Ranging Service: An Analysis. By Matt Wulff. Westminster, MD: Heritage Books, 2007.

Ronin, S.Sgt. Chris USMC (Ret.) (former Marine scout/sniper and special operator). In conversations with author from 2017 to 2019.

Ross, Bill D. *Iwo Jima: Legacy of Valor*. New York: Vintage, 1986.

Sergeant of South African descent from U.S. Marine tank battalion scout platoon. In e-mail to author on 20 May 2017.

Shaw, John and the editors of Time-Life Books. *Red Army Resurgent.* Chicago: Time-Life Books, 1979.

Six Strategies for War: The Practice of Effective Leadership. Translated by Alan Chong. Singapore: Asiapac Books, 1993.

Soard, Charles (experienced infantry scout and frequent visitor to Vietnam. In telephone conversation with author on 28 March 2001.

Spanish American War website. From this url: https://www.spanamwar.com

The Spearhead (5th Marine Division's unofficial website). From the following url: http://www.5thmarinedivision.com.

The Strategic Advantage: Sun Zi and Western Approaches to War. Edited by Cao Shan. Beijing: New World Press, 1997.

Sun Bin's Art of War: World's Greatest Military Treatise. Translated by Sui Yun. Singapore: Chung Printing, 1999.

Sun Tzu. *The Art of War.* Translated by Thomas Cleary. Boston: Shambhala Publications, 1988.

Sun Tzu. *The Art of War.* Translated and with an introduction by Samuel B. Griffith. Foreword by B.H. Liddell Hart. New York: Oxford University Press, 1963.

Super-Squad: The Now Missing Component. By H. John Poole. Emerald Isle, NC: Posterity Press, 2020.

Tactics of the Crescent Moon: Militant Muslim Combat Methods. By H. John Poole. Emerald Isle, NC: Posterity Press, 2004.

Terrorist Trail: Backtracking the Foreign Fighter. By H. John Poole. Emerald Isle, NC: Posterity Press, 2006.

Thai, Gen. Hoang Van. *How South Vietnam Was Liberated.* Hanoi: The Gioi Publishers, 1996.

Their Time in Hell: The 4th Marine Brigade at Belleau Wood. By George B. Clark. Pike, NH: The Brass Hat, 1996.

This Kind of War. By T.R. Fehrenbach. Washington, D.C.: Brassey's, 1994. Originally published in New York by MacMillan in 1963.

The Tiger's Way: A U.S. Private's Best Chance for Survival. By H. John Poole. Emerald Isle, NC: Posterity Press, 2003.

Trudeau, Noah Andre. *The Last Citadel.* Boston, MA: Little, Brown & Co., 1991.

Truong Chinh. *Primer for Revolt.* Introduction by Bernard B. Fall. New York: Praeger, 1963.

Verstappen, Stefan H. *The Thirty-Six Strategies of Ancient China.* San Francisco: China Books & Periodicals, 1999.

Wikipedia Encyclopedia. From its website, https://en.wikipedia.org.

The Wiles of War. Composed and translated by Sun Haichen. Beijing: Foreign Languages Press, 1991.

World History Encyclopedia. From its website, www.worldhistory.org.

Wukovits, John. *American Commando.* New York: New American Library, 2009.

Xiaobing Li. *China's Battle for Korea: The 1951 Spring Offensive.* Bloomington, IN: Indiana University Press, 2014.

Yahara, Hiromichi. *The Battle for Okinawa*. Translated by Roger Pineau and Masatoshi Uehara. New York: John Wiley & Sons, 1995.

"Yemen's Houthis and the Expansion of Iran's Axis of Resistance." By Katherine Zimmerman, American Enterprise Institute, March 2022.

You Are Worth It: Building a Life Worth Fighting For. By Kyle Carpenter, Don Yaeger, et al. N.P.: HarperAudio, 2019.

Yousaf, Brigadier Mohammad and Maj. Mark Adkin. *Bear Trap: Afghanistan's Untold Story*. South Yorkshire, UK: Leo Cooper, n.d. From www.afghanbooks.com.

Yousaf, Brigadier Mohammad. *The Silent Soldier: The Man Behind the Afghan Jehad*. South Yorkshire, UK: Leo Cooper, n.d. From www.afghanbooks.com.

Zabih, Sepehr. *The Iranian Military in Revolution and War*. London: Routledge, 1988.

Videotapes, Movies, TV Programs, Slide Shows, DVDs, CDs, Illustrations

"The Battle of Biak: 'A Terrifying Glimpse Into the Soul of Mankind.'" By David Alan Johnson. *Warfare History Network*, n.d. As retrieved in September 2020 from the following url: https://warfarehistorynetwork.com/2018/12/30/the-battle-of-biak-a-terrifying-glimpse-into-the-soul-of-mankind/.

"Battle of Jackson." Civil War: Secret Missions. *History Channel*, 15 January 2022. As retrieved in May 2022 from the following url: https://www.youtube.com/watch?v=wJaH3GlhN2Q.

"The Battle of the Minds." Segment of *Korea—the Unknown War Series*. London: Thames TV in association with WGBH Boston, 1990. NC Public TV, n.d.

"Gunsmoke." *Greenville Channel 7.3 (WITN)*, 23 July 2022.

"How a German Squad Attacks a Position (WWII)." *Footage Archive*. From a training film entitled "The German Infantry Squad in Action: A Demonstration of Minor Field Tactics." As retrieved in December 2018 from this url: www,youtube,com/watch?v=GDZMJXaADQI.

"Jack Lucas and the Battle of Iwo Jima." *The History Guy*. Youtube video presentation, 26 February 2020. As retrieved in February 2020 from the following url: https://www.youtube.com/watch?v=JKM0OiQZe4g.

Korea—the Unknown War. London: Thames TV in assoc. with WGBH Boston, 1990. NC Public TV.

"Last Days of World War II." *History Channel*, 28 March 1999.

"Mail Call." *DEFY TV*, 18 October 2022.

Photograph caption of the An Hoa Explosion. *Pinterest,* n.d. As retrieved in April 2017 from this url: https://www.pinterest.com/pin./347480927472222308/.
"Platoon." Movie. Written and directed by Oliver Stone. Hemdale Film Corporation, 1986.
"Return of the Taliban," *PBS's Frontline.* NC Public TV, 8 January 2007.
Sands of Iwo Jima. Videocassette. Hollywood: Republic Pictures, 1988.
Saving Private Ryan. DVD. Paramount Pictures, n.d.
"Scotland." Rick Steve's Europe. *PBS's Explorer Channel,* July 2022.

Newspaper, Magazine, and Website Articles

"About the Joint Chiefs of Staff." Www.jcs.mil, August 2022.
"American Indian Wars." *History Channel,* 17 November 2019. As retrieved in May 2022 from the following url: https://www.history.com/topics/colonial-america/american-indian-wars-timeline.
"An Hoa Project." By Warren Stansbury, n.d. As retrieved in April 2017 from the following url: http://anhoaproject.co.uk/story-warren-stansbury.html.
"Army Chain of Command," Louisiana State University Army ROTC. As retrieved in June 2022 from the following url: https://www.lsu.edu/hss/milsci/cadre/army_chain_of_command.php.
"Art of Manliness Podcast #526." Interview of Sam Gwynne by Brett, of 17 July 2017. As retrieved in March 2022 from the following url: https://www. artofmanliness.com/articles/sc-gwynne-interview-comanches/.
"Attacks on MAG-16, Hill 22, and the Attempted Attack on the Da Nang Airfield." *3rd Marines in Vietnam* website (www.3rdmarines.net), n.d.
"Backgrounder on Afghanistan: History of the War," as extracted from www.geocities.com on 29 July 2004.
"The Battle at Clark Fork." United States Department of Agriculture Forest Service, n.d. As retrieved in March of 2022 from the following url: https://www.fs.usda.gov/detail/npnht/learningcenter/history-culture/?cid=fsbdev3_055700.
"Battle Hymn of the Republic: The Story Behind the Song." The Kennedy Center. As retrieved in May 2022 from the following url: https://www.kennedy-center.org/education/resources-for-educators/classroom-resources/media-and-interactives/media/music/story-behind-the-song/the-story-behind-the-song/the-battle-hymn-of-the-republic/.

"Battle of Bunker Hill." *History Channel,* n.d. As retrieved in April 2022 from the following url: https://www.history.com/topics/american-revolution/battle-of-bunker-hill.

"Battle of Peleliu" *Facts and Details,* n.d., As retrieved in December 2020 from the following url: http://factsanddetails.com/asian/ca67/sub428/entry-5336.html.

"Battle of San Juan Hill." By Raymond K. Bluhm. *Encyclopedia Britannica.* As retrieved in May 2022 from the following url: https://www.britannica.com/event/Battleof-San-Juan-Hill.

"The Battle of San Juan Hill and El Caney." By SGM Chris Kalili, SGM Brian Pierce, and MSG Linda Williams. Probable Sergeant Major Academy thesis. As retrieved in May 2022 from this url: https://apps.dtic.mil/sti/pdfs/AD1142585.pdf.

"Belleau Wood: Marines' Shrine." By Linda D. Kozaryn, Armed Forces Press Service. *Leatherneck Magazine*, 9 August 1998.

"Biên Hòa Air Base Ammo Dump Explosion Sapper Attack." By Arnold John Houchin. Originally written on 12 January 1972, then published by the Assoc. for USAF Vietnam War Veterans of Air/Security Police, through *Vietnam War Stories,* in 2005.

"Black Jack in Cuba: General John J. Pershing's Service in the Spanish-American War." By Kevin Hymel. U.S. Army Center of Military History. As retrieved in May 2022 from the following url: https://history.army.mil/documents/spanam/WS-Prshg.htm.

Burke, Jason. "Waiting for a Last Battle with the Taliban." *The Observer* (UK), 27 June 1999.

Burns, Robert (AP). "Afghan Progress Is Cited." *Jacksonville Daily News* (NC).

"Challenges to Iran's Role in Iraq in the Post-Soleimani Era." By Hamidreza Azizi, Stiftung Wissenschaft und Politik, 22 July 2021. As retrieved in August 2022 from the following url: https://www.swp-berlin.org/en/publication/challenges-to-irans-role-in-iraq-in-the-post-soleimani-era

"Cheyenne War Parties and Battle Tactics." *Colorado History* on line, n.d. As retrieved in March 2022 from the following url: http://mr_sedivy.tripod.com/colorado6.html.

"Cochise and the Bascom Affair." By Jay W. Sharp, *DesertUSA.com.* As retrieved in February 2023 from the followiing url: https://www.desertusa.com/ind1/Cochise.html.

"The Comanche." *Visit Big Spring Texas*, n.d. As retrieved in March 2022 from the following url: http://visitbigspring.com/211/The-Comanche.

"Confucianism and Shintoism—Similarities and Differences in Belief and Practice." By Nixie Adams. *Interfaith Now,* 16 Sept. 2021. As retrieved in September 2022 from the following url: https://medium.com/interfaith-now/confucianism-and-shint%C5%8Dism-similarities-and-differences-in-belief-and-practice-7d796aea54a1.

Cooney, Daniel (AP), "Rocket Attacks Kills 2 Americans at Base in Afghanistan." *Jacksonville Daily News* (NC), 9 June 2005.

Cooney, Daniel (AP), "U.S. Kills 12 Insurgents in Afghanistan." *Jacksonville Daily News* (NC), 23 May 2005.

Cooney, Daniel (AP), "Violence Spikes in Afghanistan." *Jacksonville Daily News* (NC), 26 July 2005.

"Counting Coup." By Dennis Gaffney. *PBS's Antiques Roadshow,* 24 April 2006. As retrieved in March 2022 from the following url: https://www.pbs.org/wgbh/roadshow/stories/articles/2006/4/24/counting-coups.

"Crumbling ruins of the Mound of the Dead: World's only Bronze Age city could vanish in just 20 years." By Tara Brady, *The Daily Mail* (UK) on line, 14 October 2013. As retrieved in March 2022 from the following url: https://www.dailymail.co.uk/sciencetech/article-2458500/Worlds-surviving-Bronze-Age-metropolis-Mohenjodaro-lost-forever-ancient-city-begins-crumble.html.

"Did Lt. Gen. Richard Ewell Lose the Battle of Gettysburg." *HistoryNet,* n.d. As retrieved in May 2022 from the following url: www.historynet.com › did-lt-gen-richard-ewell-lose-the-battle-of-gettysburg.

"Discipline and Training of Soldiers." Civil War Leadership blog. As retrieved in May 2022 from the following url: http://civil-war-officers.leadr.msu.edu/2015/04/15/discipline-training-of-soldiers-2/.

"Everything You Know About the Indian Wars Is Wrong." *HistoryNet,* n.d. As retrieved in March 2022 from the following url: https://www.historynet.com/everything-know-indian-wars-wrong.htm.

"The fall of Kabul: a 20-year mission collapses in a single day." By Emma Graham-Harrison and Luke Harding. *The Guardian* (UK) on line, 15 August 2021. As retrieved in August 2022 from the following url: https://www.theguardian.com/world/2021/aug/15/the-fall-of-kabul-a-20-year-mission-collapses-in-a-single-day.

"Forward Looking Infrared Radar." *Fandom*, n.d. As retrieved in September 2022 from the following url: https://vietnamwar.fandom.com/wiki/Forward_looking_infrared_radar.

"From Amiens to Armistice: The Hundred Days Offensive." Imperial War Museums, n.d. As retrieved in June 2022 from the following url: https://www.iwm.org.uk/history/from-amiens-to-armistice-the-hundred-days-offensive.

"From Smooth Bore to Rifled Musket." *HistoryMaking.org.* As retrieved in January 2023 from the following url: http://historymaking.org/textbook/items/show/275.

Fry, J.C. "Battle Drill." *Combat Forces Journal,* April 1953. In "Development of the Squad," by Hashim, p. 27.

"Gen. Erskine quotes." As retrieved in August 2022 from this url: https://loveexpands.com/author/graves-b-erskine/.

"General Seeks New Afghan Approach." *Philadelphia Inquirer,* 1 August 2009.

"German Infantry Anti-Tank Tactics 1941/1942: Eastern Front Edition," *Military History Visualized,* n.d.

"Getting American Soldiers Trained in the Realities of Trench Warfare." America Battle Monuments Commission, 1 Nov. 2017. As retrieved in May 2022 from the following url: https://www.abmc.gov/news-events/news/getting-american-soldiers-trained-realities-trench-warfare.

Giap, Vo Nguyen. As quoted in "The Tiger and the Elephant: Viet Minh Strategy and Tactics." As retrieved in March 2022 from the following url: http://indochine54.free.fr/vm/tiger.html.

Giap, Gen. Vo Nguyen. "Once Again We Will Win." As quoted in *The Military Art of People's War,* edited by Russel Stetler. New York: *Monthly Review Press,* 1970.

Holahan, David. Review of *Empire of the Summer Moon.* As retrieved in March 2022 from the following url: https://davidholahan.com/bookreviews/empire-of-the-summer-moon.

"Honesty: The Importance Of Truth In Native American Culture." *IPL.org.* As retrieved in February 2023 from the following url: https://www.ipl.org/essay/Honesty-The-Importance-Of-Truth-In-Native-FJE5HNCTYV.

"How Alexander Hamilton's Men Surprised the Enemy at the Battle of Yorktown." By Leslie Kennedy. *History Channel,* 19 October 2021. As retrieved in April 2022 from the following url: https://www.history.com/news/alexander-hamilton-battle-yorktown-revolutionary-war.

"How the U.S. Military Learned to Learn in World War I: Lessons from the American Expeditionary Forces." By Michael Shurk, Rand Corporation. *WarontheRocks.com,* 22 August 2018.

"How wide and long would a column be." *Civilwartalk.com.* As retrieved in February 2023 from the following url: https://civilwartalk.com/threads/how-wide-and-long-would-a-column-be-and-what-did-they-consider-mechanics.136946/,

"An Indian Fight: Battle Between the Sioux and Chippewas," St. Paul Pioneer (MN), 12 July 1861. As retrieved in March 2022 from the following url: https://mail.aol.com/b/folders/1/messages/ANf5KFM-gS76YikGGww1ALRzzz8?.lang=enus&folderType=INBOX&showImages=true&offset=0&unblockNow=false.

"Indians Don't Attack at Night." *Jeff Arnold's West,* 26 June 2020. As retrieved in March 2022 from the following url: http://jeffarnold blog.blogspot.com/2020/06/indians-dont-attack-at-night.html.

"In Normandy, American Ingenuity Defeated the Impenetrable." By Herb Ditzel III. *The New York Times* on line, 5 June 1994.

"Iwo Jima and Okinawa: Death at Japan's Doorstep." *National WWII Museum* website, 10 July 2017.

Jelinek, Pauline (AP). "U.S. Plans Spring Offensive in Afghanistan." *AOL News,* 30 January 2004.

Kennedy, President John F. Quote from his 1962 West Point graduation speech. In "Unconventional Warfare Fundamentals," at *The Irregular Warrior* website.

Khan, Noor (AP), "Afghan Cops Abducted by Taliban." *Jacksonville Daily News* (NC), 19 June 2004.

Khan, Noor (AP), "Four American Soldiers Hurt in Afghanistan Blast." *Jacksonville Daily News* (NC), 14 June 2005.

"The Korean War Atrocities No One Wants to Talk About." By Jim Bovard. *The American Conservative,* 26 June 2020. As retrieved in July 2022 from this url: https://www.theamericanconservative.com/the-korean-war-atrocities-no-one-wants-to-talk-about/.

"Last Days of the Comanches." Texas Monthly on line, n,d. As retrieved in March 2022 from the following url: https://www.texasmonthly.com/articles/last-days-of-the-comanches/.

"Lost Battles of the Vietnam War." As retrieved in April 2017 from the following url: http://www.g2mil.com/lost_vietnam.htm.

"Lyndon B, Johnson: Foreign Affairs." By Kent Germany. University of Virginia Miller Center. As retrieved in July 2022 from this url: https://millercenter.org/president/lbjohnson/foreign-affairs.

"Map May Hold Key." By David. S. Cloud. *Los Angeles Times* on line, 3 May 2015.

McDowell, Patrick (AP). "Pakistan Goes After Terrorists on Border." *Jacksonville Daily News* (NC), 9 January 2004.

McGirk, Tim. "Battle in 'the Evilest Place'." *Time Magazine,* 3 November 2003, East Asia edition.

Meixler, Louis (AP). "Extremists Send Recruits into Iraq via Iranian Border." *Jacksonville Daily News* (NC), 8 November 2004.

"Memories of Massacres Were Long Suppressed Here: Tourists Now Retrace the Atrocities." By Choe Sang-Hun (South Korea Dispatch). *New York Times* on line. Asia Pacific Edition, 28 May 2019. As retrieved in July from this url: https://www.nytimes.com/2019/05/28/world/asia/south-korea-jeju-massacres.html.

"Meuse-Argonne," Library of Congress, n.d. As retrieved in June 2022 from this url: https://www.loc.gov/exhibitions/world-war-i-american-experiences/about-this-exhibition/over-there/meuse-argonne/.

"Military Drills of the Civil War." *Civilwar.com,* n.d. As retrieved in May 2022 from the following url: 8547-military-drills-of-the-civil-war.html.

"Military: Iran Teaching Iraqi Militia to Build Bombs." From Associated Press. *Jacksonville Daily News* (NC), 12 April 2007.

"Model 1893 Spanish Mauser," *Military Guns Australia,* n.d. As retrieved in May 2022 from the following url: https://firearms.net.au/military/index.php?option=com_content&view=article&id=52&Itemid=76.

"Nez Perce & U.S. Cavalry." United States Department of Agriculture Forest Service, n.d. As retrieved in March of 2022 from the following url: www.fs.usda.gov/detail/npnht/learningcenter/history-culture.

"Nez Perce Fight Battle of Big Hole," *History.com,* n.d. As retreived in March 2022 from the following url: https://www.history.com/this-day-in-history/nez-perce-fight-battle-of-big-hole.

"Night Attack – The Forsaken Tactic." *Craigchapman.com,* 31 October, 2017.

"No Word for Lying [in Apache]." Language Log. As retrieved in February 2023 from the following url: https://languagelog.ldc.upenn.edu/nll/?p=1089.

"Ottowa Indians." *Ohio History Central,* n.d. As retrieved in March 2022 from the following url: https://ohiohistorycentral.org/w/Ottawa_Indians.

"Overstressing Safety Is Leaving Momentum Behind." By Lt.Col. H.J. Poole USMC (Ret.). *Army Magazine,* August 2020.

"Overview of the American Revolutionary War." *American Battlefield Trust.* As retrieved in April 2022 from the following url: https://www.battlefields.org/learn/articles/overview-american-revolutionary-war?ms=googlegrant&ms=googlegrant.

"Papal Encyclical Urges Capitalism to Shed Injustices." By Peter Steinfels. *New York Times* on line, 3 May 1991. As retrieved in October 2022 from this url: https://www.nytimes.com/1991/05/03/world/papal-encyclical-urges-capitalism-to-shed-injustices.html.

Peter, Tom A. "U.S. Soldiers: Iraq, This Isn't." *Christian Science Monitor,* 29 November 2009.

Peterson, Scott. "Marine, Insurgent Tactics Evolve." *Christian Science Monitor,* 17 November 2004.

Peterson, Scott. "Shiites Rising—Part One." *Christian Science Monitor,* 6 June 2007.

"Place of the Loyal Samurai." By Daniel Weiss. *Archaeology Magazine*, July/August 2019.

"Quana Parker . . . ." *The Oklahoman* on line, 20 July 2002. As retrieved in March 2022 from the following url: https://www.oklahoman.com/article/2002720/quanah-parker-maybe-not-a-wonderful-person-but-truly-a-great-man.

"Remembering Okinawa." By Robert Akers Martin. *Lulu.com,* 2014.

"The Rifle-Musket vs. The Smoothbore Musket, a Comparison of the Effectiveness of the Two Types of Weapons Primarily at Short Ranges." By Justin Stanage. Communicated by: Dr. Donald B. Marty, Department of History, Indiana Universtiy, n.d. As retrieved in February 2023 from url: file:///Users/henrypoole/Downloads/19841-Article%20Text-43984-1-1020150825%20(1).pdf.

"The Rise and Fall of the Comanche Empire." *NPR,* 20 May 2011. As retrieved in March 2022 from the following url: https://www.npr.org/2011/05/20/136438816/the-rise-and-fall-of-the-comanche-empire.

"The Rough Riders Storm San Juan Hill, 1898." *Eyewitness to History,* 2004. As retrieved in May 2022 from the following url: http://www.eyewitnesstohistory.com/roughriders.htm.

"A Season in Hell: Iwo Jima 50th Anniversary." By Nathan Miller. *Baltimore Sun,* 19 February 1995. As retrieved in February 2020 from the following url: https://www.baltimoresun.com/news/bs-xpm-1995-02-19-1995050098-story.html.

"Secret Manual Gives Glimpse of North Korean Military Tactics." By Steve Herman. *Voice of America* on line, 18 September 2010.

Segal, Naomi (Jewish Telegraphic Agency). "IDF Absolved of Blame in Deaths of Naval Commandos in Lebanon." *San Francisco Jewish Community Publication,* 31 October 1997.

"The Seminole Wars." Florida Dept. of State, n.d. As retrieved in March 2022 from the following url: https://dos.myflorida.com/florida-facts/florida-history/seminole-history/the-seminole-wars/.

"Setting the Record Straight About Native Languages: Ishmay Ohmay ." Native-Languages.org. As retrieved in February 2023 from the following url: http://www.native-languages.org/iaq24.htm.

"71st Regiment Infantry New York Volunteers Spanish-American War." New York Museum and Veterans Research Center. As retrieved in May 2022 from the following url: https://museum.dmna.ny.gov/unit-history/conflict/spanish-american-war-1898/71st-regiment-infantry.

"76 Militants Die in Afghan Battles." World Brief Wire Reports (AP). *Jacksonville Daily News* (NC), 23 June 2004.

"Significant Civil War Battles." *PBS's American Experience,* n.d. As retrieved in May 2022 from the following url: https://www.pbs.

"Small Arms of the Civil War." *American Battlefield Trust.* As retrieved in January 2023 from the following url: www.battlefields.org › learn › articles › small-arms-civil-war.

"The Spanish Modelo 1893 Mauser Rifle." *Shooting Times* on line, 29 March 2017. As retrieved in May 2022 from the following url: https://www.shootingtimes.com/editorial/longgun_reviews_st_spanmodelo1892mr_200812/100180. Access denied within one hour of original retrieval.

"The Spanish 7x57mm Mauser," Spanish American War website. As retrieved in May 2022 from the following url: https://www.spanamwar.com/spanishmauser.htm.

Stevenson summary of "War on the Run: The Epic Story of Robert Rogers and the Conquest of America's First Frontier." By John Ross. *Tactics and Preparedness,* November 2021.

Stoy, Lt.Col. Tim U.S.Army (Ret.),. Review of *Okinawan Reckoning.* By H, John Poole. Emerald Isle, NC: Posterity Press, 2022. *Army Magazine,* January 2023.

Straziuso, Jason (AP). "Afghan Bomber Deadlier." *Jacksonville Daily News* (NC), 3 June 2007.

"Suicide Bomber Kills Peacekeeper, Civilian in Afghan Capital." World Brief Wire Reports (AP). *Jacksonville Daily News* (NC), 28 January 2004.

Surkov, Gen. Mikhail. As quoted in "The World Turned Upside Down: Military Lessons of the Chechen War." By Anatol Lieven. *Armed Forces Journal International,* August 1998.

"Syria's Civil War Produces a Clear Winner: Hezbollah," By Maria Abi-Habib. *Wall Street Journal* on line, 3 April 2017.

"Tactics: Rethinking the Basis of Infantry Close Combat." By Lt.Col. David Kilcullen. *Australian Infantry Journal*, 2 April 2015.

Taylor, Vic "Hotel Company-Day Three." *Marine Corps Gazette,* April 2004.

"These four final battles spelled the end of Comanche hegemony over the Southern Plains." By Richard A. Marini. San Antonio: *San Antonio Express News,* 12 April 2017. As retrieved in March 2022 from the following url: https://www.expressnews.com/news/local/article/These-four-final-battles-spelled-the-end-of-11069732.php.

"This Day in History, October 7." *History Channel*, n.d. As retrieved in August 2022 from the following url: https://www.history.com/this-day-in-history/u-s-led-attack-on-afghanistan-begins.

"This Day in History September 26." *History Channel*, n.d. As retrieved in June 2022 from the following url: https://www.history.com/this-day-in-history/meuse-argonne-offensive-opens.

"This Day in History, September 26, 1918: The Meuse-Argonne Campaign Begins." America Battle Monuments Commission, n.d. As retrieved in June 2022 from the following url: https://www.abmc.gov/multimedia/videos/day-history-september-26-1918-meuse-argonne-campaign-begins.

"This Day in History September 29." *History Channel,* n.d. As retrieved in June 2022 from the following url: https://www.history.com/this-day-in-history/allied-forces-break-through-the-hindenburg-line.

"This Old Gun: Model 1893 Spanish Mauser." By Garry James. *American Rifleman,* 17 February 2002. As retrieved in May 2022 from the following url: https://www.americanrifleman.org/content/this-old-gun-model-1893-spanish-mauser/.

Tran, Tini (AP). "Marines, Iraqis Launch Offensive." *Jacksonville Daily News* (NC), 24 November 2004.

"The 22nd U.S. Infantry in the Spanish-American War," Spanish American War website. As retrieved in May 2022 from the following url: https://www.spanamwar.com/22ndus.htm.

"Two Insurgents Killed in Afghanistan Fight." World Briefs Wire Reports (AP). *Jacksonville Daily News* (NC), 25 May 2005.

Tyson, Ann Scott. "Going in Small in Afghanistan." *Christian Science Monitor,* 14 January 2004.

"Ulysses S. Grant's Overland Campaign: Six Bloody Weeks." By Christopher Klein. *History Channel,* updated 21 April 2020. As retrieved in May 2022 from the following url: https://www.history.com/news/grants-overland-campaign-civil-war.

"The U.S. deal with the Taliban destroyed Afghans' military morale." By Bill Chappell. *NPR* on line, 18 May 2022. As retrieved in August 2022 from the following url: https://www.npr.org/2022/05/18/1099688825/sigar-afghanistan-forces-collapse.

"U.S. Soldier, 7 Militants Die in Afghan Firefight." World Briefs Wire Reports (AP). *Jacksonville Daily News* (NC), 11 June 2005.

"The War Effort: The War Department." Mrlincolnswhitehouse.org. As retrieved in May 2022 from the following url: http://www.mrlincolnswhitehouse.org/washington/the-war-effort/war-effort-war-department/.

Wazir, Ansnullah (AP). "Fight Wanes between Troops, Suspected al-Qaida." *Jacksonville Daily News* (NC), 22 March 2004.

"Westmoreland." *History Channel* on line, 29 May 2019. As retrieved in July 2022 from the following url: https://www.history.com/topics/vietnam-war/william-westmoreland.

"What did men do in the army." Bitesize Topics. *BBC* on line, n.d. As retrieved in May 2022 from the following url: https://www.bbc.co.uk/bitesize/topics/zqhyb9q/articles/zkqtrj6.

"What Kind of Techniques Did the French Use to Fight in WWI," By Ralph Heibutzki. *Seattle Post-Intelligencer* on line, n.d. As retrieved in May 2022 from the following url: https://education.seattlepi.com/kind-techniques-did-french-use-fight-wwi-5815.html.

"What Was the Cultural Revolution." By Kallie Szczepanski. *About.com,* n.d. As retrieved in October 2014 from the following url: http://asianhistory.about.com/od/modernchina/f/What-Was-The-Cultural-Revolution.htm.

"The World of 1898: The Spanish-American War." Library of Congress Researchers, n.d. As retrieved in May 2022 from the following url: https://loc.gov/rr/hispanic/1898/intro.html.

"World War I: New Weapons, Same Old (Bad) Tactics." By Michael Peck. *National Interest,* n.d. As retrieved in May 2022 from the following url: https://nationalinterest.org/blog/reboot/world-war-i-new-weapons-same-old-bad-tactics-173462.

About the Author

After 28 years of commissioned and noncommissioned infantry service, John Poole retired from the United States Marine Corps in April 1993. While on active duty, he studied small-unit tactics for nine years: (1) six months at the Basic School in Quantico (1966); (2) seven months as a rifle platoon commander in Vietnam (1966-67); (3) three months as a rifle company commander at Camp Pendleton (1967); (4) five months as a regimental headquarters company (and camp) commander in Vietnam (1968); (5) eight months as a rifle company commander in Vietnam (1968-69); (6) five and a half years as an instructor with the Advanced Infantry Training Company (AITC) at Camp Lejeune (1986-92); and (7) one year as the Staff Noncommissioned Officer in Charge (SNCOIC) of the 3rd Marine Division Combat Squad Leaders Course (CSLC) on Okinawa (1992-93).

While at AITC, he developed, taught, and refined courses on maneuver warfare, land navigation, fire support coordination, call for fire, adjust fire, close air support, M203 grenade launcher, movement to contact, daylight attack, night attack, infiltration, defense, offensive Military Operations in Urban Terrain (MOUT), defensive MOUT, Nuclear/Biological/Chemical (NBC) defense, and leadership. While at CSLC, he further refined the same periods of instruction and developed others on patrolling.

He has completed all of the correspondence school requirements for the Marine Corps Command and Staff College, Naval War College (1,000-hour curriculum), and Marine Corps Warfighting Skills Program. He is a graduate of the Camp Lejeune Instructional Management Course, the 2nd Marine Division Skill Leaders in Advanced Marksmanship (SLAM) Course, and the East-Coast School of Infantry Platoon Sergeants' Course.

In the 25 years since retirement, Poole has researched the small-unit tactics of other nations and written 18 other books: (1) *The Last Hundred Yards,* a squad combat study based on the consensus opinions of 1,200 formal school NCOs and casualty statistics of AITC and CSLC field trials; (2) *One More Bridge to Cross,* a treatise on enemy proficiency at short range and how to match it; (3) *Phantom Soldier,* an in-depth look at the highly deceptive Asian style of war; (4) *The Tiger's Way,* the fighting styles of Eastern fireteams and soldiers; (5) *Tactics of the Crescent Moon,* insurgent procedures in Palestine, Chechnya, Afghanistan, and Iraq; (6) *Militant Tricks,* how Muslim insurgents fight in an urban environment; (7) *Terrorist Trail,* tracing the *jihadists* in Iraq back to their home countries; (8) *Dragon Days,* an unconventional warfare technique manual; (9) *Tequila Junction,* how to fight narco-guerrillas; (10) *Homeland Siege,* countering the 4GW assault on America by a foreign power's organized-crime proxies; (11) *Expeditionary Eagles,* how to outmaneuver the Taliban; (12) *Global Warrior,* forestalling WWIII through tiny-contingent activity; (13) *Gung Ho,* how supporting arms are not needed to take strongpoint matrices; (14) *Strategic Rifleman,* how to produce tiny semi-autonomous combat elements; (15) *Afrique,* with Africa as a model how U.S. intelligence is being gathered incorrectly for certain adversaries; (16) *Sinoland,* how the U.S. has been under 4GW attack from China since 9/11; (17) *Guardian Joe,* how—through more skilled infantry squads—the Pentagon can use less force to more easily win a modern war; (18) *North Korea's Hidden Assets,* 13 little-known wartime capabilities of what has been erroneously called the Hermit Kingdom; (19) *Super-Squad: The Now Missing Component;* (20) *The Iwo Alamo;* (21) *Peleliu Progress;* and *(22) Okinawan Reckoning.*

Since retirement, Poole has traveled extensively in both Communist and Islamist worlds to study the infantry tactics of America's enemies. He has done research in Mainland China (twice), North Korea, Vietnam, Cambodia, Pakistan (twice), Bangladesh, Malaysia, Iran, Lebanon, Turkey, the Emirates, Egypt, Sudan, Tanzania, Zambia, South Africa, Morocco, and Venezuela. Over the course of his lifetime, he has visited scores of other nations on all five continents. As of April 2014, John Poole had conducted multi-day training sessions (on advanced squad tactics) at 41 (mostly Marine) battalions, nine Marine schools, and seven special-operations units from all four U.S. service branches.

Between early tours in the Marine Corps (from 1969 to 1971), he served as a criminal investigator with the Illinois Bureau of Investigation (IBI). After attending the State Police Academy for several months in Springfield, he was assigned to the IBI's Chicago office. There, he worked mostly on general criminal and drug cases.

Several periods in Poole's life have qualified him to accurately mirror the viewpoints of America's true fighters— its junior enlisted infantrymen. In 1965, his first official orders were for "Private Poole." From 1992 to 1993, he served as SNCOIC of a divisional squad leaders school. Then, almost all of the 1100 students who attended his "post-retirement" training sessions were E-5's and below.

Poole is one of the very few Americans with an extensive enough background to know how to fix the ongoing tactical shortfall within the U.S. military. After some 60 multi-day training sessions at different active-duty units, he has developed, tested, and refined its solution (through a "bottom-up" training supplement). He has also become adept at the now virtually lost art of field training light infantrymen (those who are able do well on their own in the absence of orders).

Name Index

A

Alligator, Chief 93
Amherst, Gen. Jeffrey 39
Appleman. Roy E. 153
al-Assad, Pres. Bashar Hafez 194, 195

B

Baldwin, Gen. Theodore A, 106
Barker, M.Gen. David B. 270
Barrow, Capt. Robert H. 166, 167
Beauregard, Gen. P, G, T. 74
Berry, Maj. Benjamin S. 127
Biden, Pres. Joseph R., Jr. 212
bin Laden, Osama 207
Birney, M.Gen. David B. 75
Black Hawk, Chief 72
Bouqet, Col. Henry 37
Braddock, M.Gen. Edward 33, 37, 38, 64
Brainard, PFC Morgan 166
Buffalo Hump, Chief 84
Burgoyne, Gen, John 61

C

Cadillac, Antoine s de la Mothe 35
Carlson, Lt.Col. Evans Fordyce 145
Chaffee 103
Champion, B.Gen. Gregory 208
Chung Il-Wan 164
Clark, Rogers 62
Clark, Lt. William 84
von Clausewitz, Karl xv
Clinton, Gen. Henry 67, 68
Cornwallis, M,Gen. Charles 63, 67, 68, 70
Cumings, Bruce 174

D

Dalyell, James 39
Davis, Lt.Col. Raymond G. 159
Davis, William C. 93
Dragging Canoe, Chief 71
Dupuy, Trevor Nevitt 156

E

Edson, Lt.Col. Merritt A. "Red Mike" 145
Eisenhower, Pres. Dwight D. 264
Engels, Friedrich 28
Ewell, Lt.Gen. Richard S. 74

F

Fehrenbach, T. R. 172, 173
Foch, Gen. Ferdinand 136
Fonda, Jane 73
Fox, Capt. Wesley L. 15
Frontinus, Sextus Julius xv
Fry, M.Gen. James C. 176

G

Gates, Gen. Horatio L. 65
Giap. Gen. Vo Nguyen 4, 95, 96, 225
Gibbon, Col. John 91
God 171, 192
Gordon, Gen. John B. 75, 76
Grant, Gen. Ulysses S. 73, 75, 76, 81, 97
Grant 86
Greene, Gen. Nathanael 63
Griffith, B.Gen. Samuel B. 66
Grimes, M.Gen. Bryan 76
Gudmundsson, Bruce I. 119

H

Hackworth, Col. David H. 178
Hamilton, Alexander 70
Haqqani, Jalaluddin 205
Hardy, Bert 164
Hatsumi, Grandmaster Masaaki (formerly Yoshiaki) 95
Hawkins, Lt.Col. Jack 165
Hekmatyar, Gulbaddin 208
Hemingway, Albert 178
Hitler, Fuhrer Adolf xvi, 155
Holcomb, Gen. Thomas 128
Honsowetz, Lt.Col. Russel E. 159
Howard, Gen. Oliver Otis 91
Howe, Gen. William 61
Huber, Thomas M. 29, 148
Hussein, Pres. Saddam 191, 196

I

Inagaki, Takeshi 29

J

Jackson, Gen, Andrew 93
Jaynes, Gregory 74
Joffre, Gen. Joseph-Jacques-Césaire 122
John Paul II, Pope 97, 264
Johnston, Gen. Joseph E. 76
Johnson, Pres. Lyndon B. 181
Jumper, Chief 93

K

Kaplan, Robert D. 202
Kaztman, Kenneth 193, 194
Kent, B.Gen. Jacob 104, 107
Khomeini, Ayatollah Ruhollah 191, 192
Kilcullin, Lt.Col. David 261, 262, 263
Kim Je-Hue 162
Knox, Henry 60
Krulak, Gen. Charles C. 126
Krulak, Gen. V. H. “Brute” 179

L

de Lafayette, Gen. Marquis 63
Lee, Charles 65
Lee, Gen. Robert E. 74, 262

Lewis, Capt. Meriwhether 84
Lewis, Sgt. 148
Lind, William S. 218
Lincoln, Pres. Abraham 72
Lincoln, Gen. Benjamin 67, 68
Lincoln, Pvt. Henry 76
Lockwood, Capt. B. C. 102
London, Private Henry 76
von Ludendorff, Gen. Erich 132, 135
Ludlow, Gen. William 101, 102, 103

M

MacDonald, Charles B. 156
Mackenzie, Col. Ranald Slidell 85, 87
Madhi, Adel Abdul 196
Maitland, Terrence 11
Mao Tse-tung 4, 11, 14, 28, 90, 92, 96, 146, 171, 172, 173, 179, 193, 224
Marion, Lt.Col. Francis 58, 65, 67, 68, 69
Marshall, Gen. George C. 156
Marshall, B.Gen S. L. A. 175, 236
Marx, Karl 28
Maslowski, Peter 156
Massoud, Ahmed Shah 203, 210
McChrystal, Gen. Stanley A. 210
McInerney, Peter 11
McPherson, James M. 75
Micanopy, Chief 93
Millet, Allan R. 156
Mitchell, Gen. William "Billy" 132
Miyamura, Cpl. Hiroshi N. 26
Montgomery, Richard 65

N

Nixon, Pres. Richard M, 182
Nolan, Keith William 52

O

Obama, Pres. Barack 211
Omar, Mullah Mohammad 203
Ord, M.Gen. Edward O. C. 75
Ord, 1stLt, Jules 106, 107
Osceola, Chief 93, 94

P

Pahlavi, Shah Mohammad Reza 192, 194
Parker, Lt. 105
Parker, Chief Quanah 85, 87, 88
Patterson, Lt.Col. 101
Patton, Gen. George S. 156, 170, 273
Pershing, Lt. John J. 105, 106, 107, 111, 112, 123, 125, 132, 134, 136
Petain, Field Marshall Henri 122
Petraeus, Gen. David H. 261
Pickens, Andrew 68
Pickett, M.Gen. George 108, 262
Pontiac. Chief 37
Puller, Lewis B. "Chesty" 51, 159, 165, 170, 171, 186, 200
Putnam, Israel 65

Q

No entries.

R

Reagan, Pres. Ronald Wilson xvi
Red Bear, Chief 86
Red Cloud, Chief 72
Redifer 272, 273
Rehman, Gen. Aktar Abdul 203
Ridge, Lt.Col. Thomas L. 165
Ridgway, Gen. Matthew 168
Rindfleisch, Bryan C. 56
Rochambeau, Gen. (Jean-Baptiste Donatien de Vimeur, comte de Rochambeau) 63
Rogers, Robert 37, 39-42, 44, 46, 47, 53, 58, 64
Rohr, Capt. Willy 119
Roosevelt. Lt.Col. Theodore 99, 104, 107, 110, 111
Rose, Bill 16
Ross, John F. 54
Ross, Richard P. 152, 159
Rupertus, Gen. William H. 270

S

Sacagawea 84
al-Sadr, Muqtada 191
Schuyler, Pbilip 65
Shafter, M.Gen. William 9, 103, 104
Sheetz, B.Gen. Joseph R. 153
Simmons, Maj.Gen. Edwin 165, 166, 168
Sledge, Cpl. Eugene B. 271, 272, 273
Smith, Gen. Holland McTyeire "Howlin' Mad" 146
Stanton, Sec. of War Edwin 79
Star, John 65
von Steuben, Baron Friedrich Wilhelm 62
Sturgis, Col. James G. 91, 92
Sulemeini, Gen. Qassem 196
Sumner, Gen, Samuel 104
Sumter, B.Gen. Thomas 68
Sun Bin 51
Sun Tzu 1, 28, 49, 70, 97, 224, 271
Surkov, Gen. Mikhail 218
Sutter, Lt.Col. Alan 165
Syngman Rhee, Pres. 164

T

Techumseh, Chief 71
Trudeau, Noah Andre 76
Trump, Pres. Donald 212

U

Upton, Col. Emery 74

V

Vandegrift, Maj.Gen. Alexander Archer 30
Van Horne, Maj. 102
Vyas, Khushaal 122

W

Walt, Gen. Lewis William 178, 179
Washington, Gen. George 38, 60, 61, 65
Westmoreland, Gen. William C. 181, 182
Wheeler 107
Wilke, Gov. Franc B. 86
Williams, Cpl. Hershel W. ("Woody") 268
Wolfe, Gen. James 39

Wukovits. John 4
Wulff, Matt 39, 40, 64

X

Xiaobing Li 172, 173

Y

Yahara, Col. Hiromichi 152
Younger, George 164
Yousaf, Brigadier Muhammad 204

Z

No entries.